8th AIChE Southwest Process Technology Conference 2016

Galveston, Texas, USA
6 – 7 October 2016

ISBN: 978-1-5108-3477-4

Printed from e-media with permission by:

Curran Associates, Inc.
57 Morehouse Lane
Red Hook, NY 12571

CURRAN ASSOCIATES INC. proceedings.com

Some format issues inherent in the e-media version may also appear in this print version.

Printed by Curran Associates, Inc. (2017)

AIChE
120 Wall Street, FL 23
New York, NY 10005-4020

Phone: (800) 242-4363
Fax: (203) 775-5177

www.aiche.org

Additional copies of this publication are available from:

Curran Associates, Inc.
57 Morehouse Lane
Red Hook, NY 12571 USA
Phone: 845-758-0400
Fax: 845-758-2633
Email: curran@proceedings.com
Web: www.proceedings.com

TABLE OF CONTENTS

The Future of Energy

Krishnamoorti, R., University of Houston

Originally presented on: 10/6/2016 8:45:00 - 9:10:00

With the world's population poised for a significant growth over the next thirty years, the future challenges of economic growth globally are closely tied to the future of energy choices and their potential impact on climate change. Disruptive technologies in efficiency, transmission and sources will form the corner stone of future choices of energy. The current state of these technologies and the life-cycle analysis of their overall greenhouse gas impact provide valuable insight into pathways for an economically and environmentally sustainable future and are the basis of the discussion.

Light Hydrocarbons Furnace Fuel Flexibility

Holmes, H.,
Bowers, D.,

Originally presented on: 10/6/2016 9:30:00 - 9:55:00

Large integrated manufacturing facilities utilize cogeneration for meeting thermal and electrical energy needs. Efficient use of by-product fuel plays a key role in optimizing site energy costs. Varying feed slates in light hydrocarbons plants necessitate new furnace fuel flexibility options to provide the best advantage for site heat rate. Working across business lines at Dow's Louisiana Operations Division results in reduced fuel use, reduced carbon emissions and operating cost for all business partners on site.

Achieving Higher Energy Efficiency in a H2 Plant via Smart Manufacturing

Kumar, A.,

Originally presented on: 10/6/2016 9:55:00 - 10:20:00

Steam methane reformers (SMRs) provide the bulk of the hydrogen consumed in refineries and for production of several important chemicals. Consequently, these are large scale systems, with modern hydrogen plants producing above 100 million standard cubic feet hydrogen per day. In this context, energy efficient operation is an imperative necessity for long-term economic viability and mathematical modeling of SMRs becomes a crucial tool for achieving high energy efficiency. The models can be either high-fidelity (e.g., using computational fluid dynamics), data-driven (and typically low-order) empirical models, or a combination of both. Furthermore, the development of mathematical models should be accompanied by online deployment. The latter must be supported by a smart manufacturing framework, that includes, i) capabilities for acquiring appropriate process data via systematically placed sensors, ii) adequate high-performance computational resources for just-in-time computations, and, iii) a user-friendly visualization interface for operator use. Moreover, rapid adoption of such practices demand that such an IT infrastructure should be easy to use and deploy. These requirements constitute the natural and essential form of a Smart manufacturing (SM). In this work, we present an example application of SM concepts to an industrial scale steam methane reformer along with the benefits obtained.

Ethylene Furnace Recovery Quench Oil Circuit Optimization

Karra, S.,
Luke, I.,
Nelson, A.,
Kemp, L.,
Gage, P.,
Originally presented on: 10/6/2016 10:20:00 - 10:45:00

A 50% reduction in heat dissipation to cooling tower water was achieved in the ethylene furnace recovery section at Beaumont Chemical Plant via improvements in constrained multi-variable control by a team of operations and engineering. Following strategy was employed successfully:

1. The Primary Fractionator cutpoint temperature was maximized against tower overheads and bottoms constraints resulted in lessening the heat losses to CT water.
2. Quench oil flow to downstream distillation tower reboilers was maximized against hydraulic constraints in quench oil circuit while minimizing supplemental reboiler steam.

This amounted to an additional 175,200 MMBTU/Y energy being recovered from quench oil as reboiler duty in downstream distillation towers while unloading the CT exchangers in QO circuit.

The Role of Steam Simulators for Advanced Controls Implementation, Operator Training, Process Safety and Plant Operational Excellence

Janvier, B., Enero Solutions

Originally presented on: 10/6/2016 10:45:00 - 11:10:00

Advanced Process Controls (APC) is usually the solution for improving steam plant runnability, safety and profitability. However, commissioning of an APC package on a steam system is risky and the long-term maintenance of these packages is so cumbersome that most APC are eventually turned off after just a few years of operation. This presentation shows how first-principle simulations can be used to reduce commissioning risk, improve the long term operability of an APC package, and train operators. The speaker will also present recent APC project results where simulators were used to test the logic, tune the controllers and train the operations team. In these projects, the "virtual steam plant" approach allowed the new APC logic to be commissioned with no production losses and with the expected long-term safety and profitability goals.

Exceptional Stability through High Performance Hydroprocessing (HPC) Catalysts

Amit Kelkar
Criterion Catalysts & Technologies

AIChE
Southwest Process Technology Conference

8th AIChE Southwest Process Technology Conference I October 6-7, 2016 I Galveston, TX

Agenda

- Function of HPC Catalysts
- Composition of HPC Catalysts
- Catalyst Technology Development
- Importance of Stability
- Examples of Activity & Stability improvements

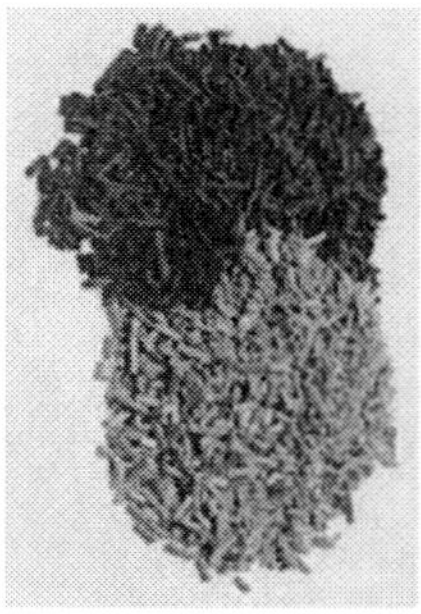

AIChE
Southwest Process Technology Conference

8th AIChE Southwest Process Technology Conference I October 6-7, 2016 I Galveston, TX

Function of HPC Catalysts

- Purification of Petroleum Fractions (Hydrotreating):
  - Removal of S & N reduces pollution from gasoline & diesel
  - Saturation of aromatics improves Smoke point of jet and Cetane quality of diesel
  - Remove contaminants and inhibitors from feeds to downstream refinery processes such as FCC, reforming, etc.
- Conversion of Higher Boiling Molecules (Hydrocracking):
  - Isomerize long straight chain paraffins into branched paraffins to improve cold flow properties
  - Crack large molecules (680 °F plus) into smaller ones to be used as fuel or Lube oil base stocks
- As the name implies, these reactions occur in the presence of H_2 at elevated temperature and pressure

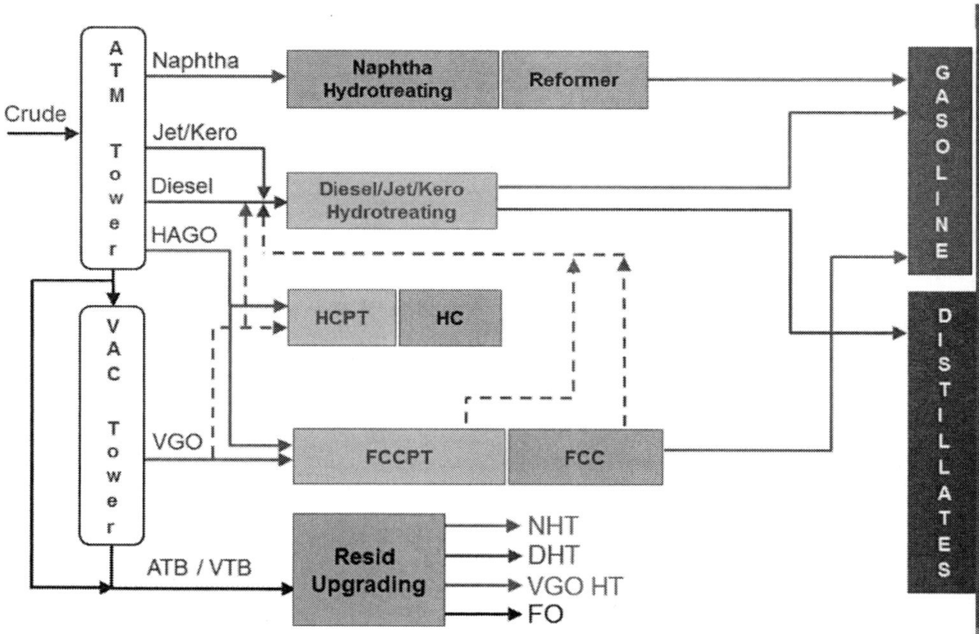

8th AIChE Southwest Process Technology Conference I October 6-7, 2016 I Galveston, TX

Composition of HPC Catalysts

- Hydrogenation Function:
  - Made up of high activity metal sulfides such as Ni, Co, Mo (base metal) or Pt, Pd (noble metal)
- Cracking Function:
  - Provided by amorphous oxides (Alumina, Silica-Alumina) or by crystalline zeolites.
  - Ratio of cracking and hydrogenation functions is critical to optimize activity and selectivity.
- Binder (Alumina):
  - High surface area enabling good dispersion of active phase
  - High void fraction to enable easy access to active sites
  - Minimal activity to prevent undesirable reactions and good strength to avoid breakage
- Hydrocracking catalysts are bifunctional in nature and include hydrogenation & cracking activity

AIChE Southwest Process Technology Conference

8th AIChE Southwest Process Technology Conference I October 6-7, 2016 I Galveston, TX

Catalyst Technology Development

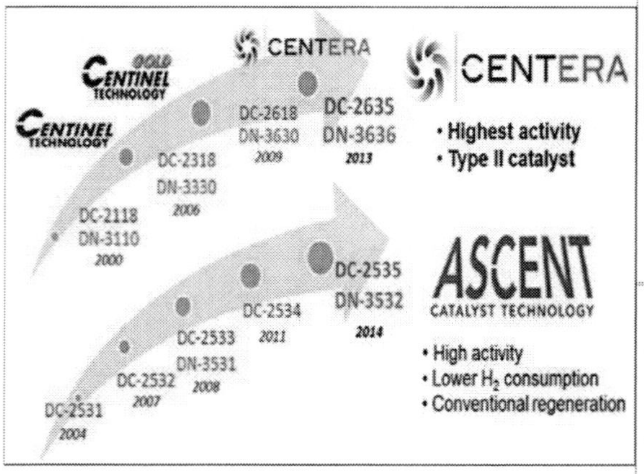

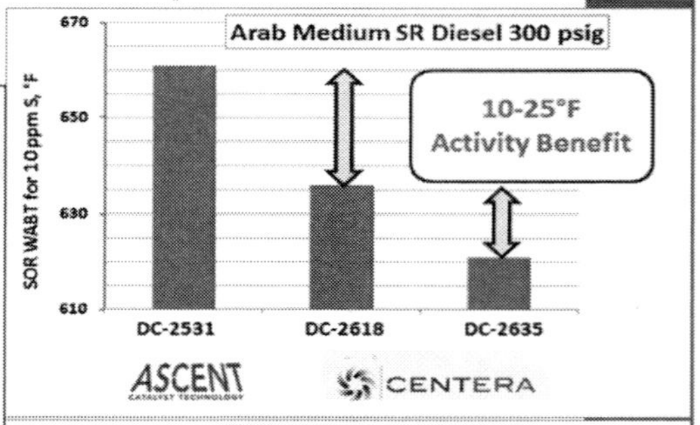

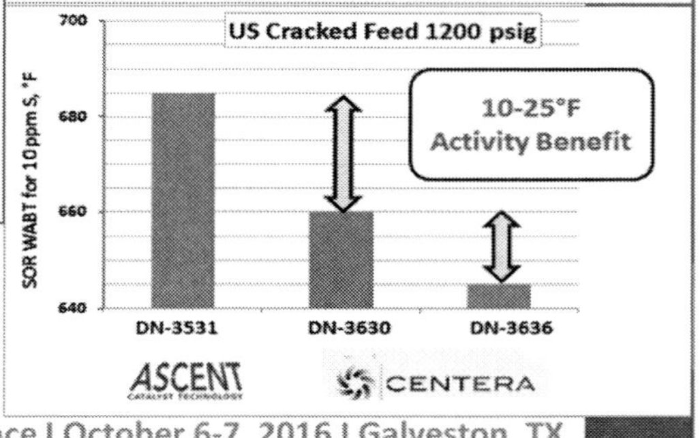

8th AIChE Southwest Process Technology Conference I October 6-7, 2016 I Galveston, TX

Importance of Stability

- While activity gain is critical in catalyst development, truly successful performance combines good activity with low deactivation rate to enable either longer life or improved ability to process more difficult feeds
- Higher hydrogenation activity means lower propensity to form coke
- Physical properties of the binder play a critical role in catalyst stability:
  - High surface area for both availability of active sites and adsorption of poisons
  - High pore volume provides storage capacity for coke & metals in case of heavier feeds

AIChE Southwest Process Technology Conference

8th AIChE Southwest Process Technology Conference I October 6-7, 2016 I Galveston, TX

Increased resilience with more cracked feed and higher severity

	ASCENT™ DN-3551	CENTERA™ DN-3651
Feed Rate, kBPD	BASE	+ 10 %
% Cracked	28 %	51 %
Feed API	BASE	- 1.9
Feed Sulfur, wt%	BASE	+ 22 %
Feed Nitrogen, ppm	BASE	+ 26 %
Feed T95, °F	BASE	BASE
Deactivation Rate, °F/m	1.2	1

	HDS Mode	Aromatic Saturation Mode
WABT, °F	BASE	+ 25 F
API Gain	BASE	BASE + 0.8
% HDS	BASE	+ 2 %
% HDN	BASE	+ 8 %
Diesel Sulfur	BASE	- 47 %
Deactivation Rate, °F/m	BASE	No change

- This FCCPT unit processed higher rate & more cracked stock at similar deactivation rate (DR) compared to previous cycle
- Operating mode was changed to maximum saturation by increasing temperature resulting in significant improvement in FCC feed quality
- The exceptional stability of the current catalyst system means the DR continues to be low even at the higher severity

AIChE Southwest Process Technology Conference

8th AIChE Southwest Process Technology Conference I October 6-7, 2016 I Galveston, TX

Excellent Stability with Increased Operating Severity

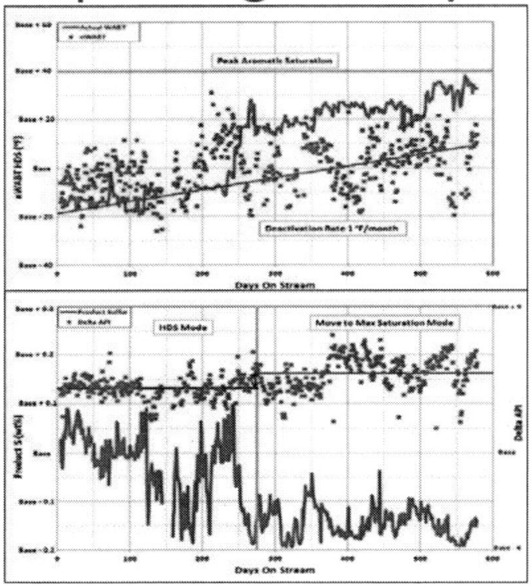

- Operating mode for this FCCPT unit was switched to maximum aromatic saturation to target higher volume swell & H_2 consumption
- Delta API was one number higher in Asat mode
- Deactivation rate was unchanged in spite of the higher severity

Stable performance with Heavier Feed in ULSD service

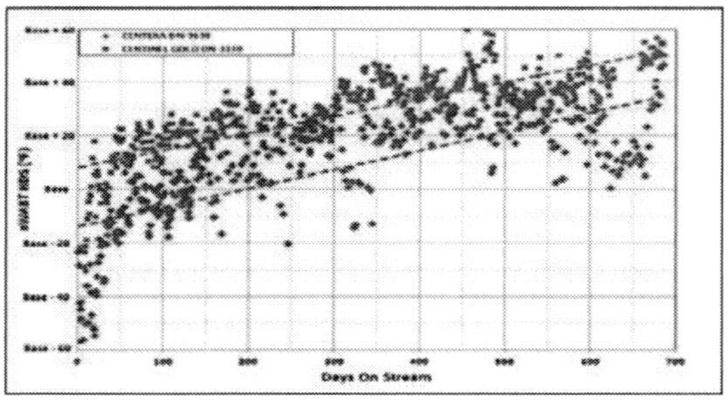

	CENTINEL ™ DN-3330	CENTERA™ DN-3630
Feed Rate, kBPD	BASE	+ 3%
Feed S, wt%	**BASE**	**+ 175%**
Feed N, ppm	BASE	+ 22%
Feed T90 / FBP, °F	**BASE / BASE**	**+ 22 / + 27 °F**

- New generation catalyst maintained a 15 – 20 °F activity benefit throughout the cycle
- Refinery was able to increase straight run diesel and LCGO end point
- Average feed was heavier with higher S and N; unit achieved higher margin while maintaining cycle life

Improved HCR PT Activity & Stability with Heavier HCGO

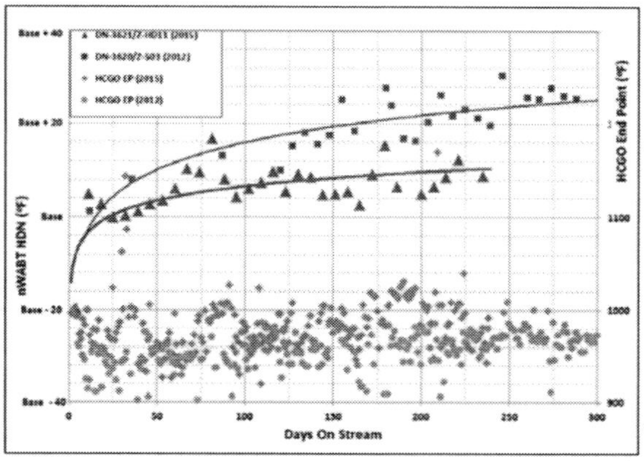

- Pretreat reactor changed to a newer generation hydrotreating and ASA catalyst system
- The current system is 8 – 10 °F higher in HDN activity with a 40% lower deactivation rate even though the HCGO end point is 20 – 40 °F higher than last cycle
- Current exotherm is 20% higher indicating increased hydrogenation

AIChE
Southwest Process Technology Conference

Excellent Stability leads to higher Diesel Yields

	ASCENT™ DN-3551 / CENTERA™ DC-2650	CENTERA™ DN-3651 / DC-2650
Feed Rate	BASE	+ 4 %
LCO Rate	**BASE**	**+ 37 %**
Feed API	BASE	- 1.0
Feed Sulfur	BASE	BASE
Feed Nitrogen	**BASE**	**+ 9 %**
H_2 Consumption	BASE	+ 11 %
Diesel Yield, % of FF	**47 %**	**52 %**
Diesel T95	**BASE**	**+ 10 °F**

- Another FCCPT unit that processes 100% cracked feed enjoyed large increase in diesel yield with a new generation catalyst load
- Unit was able to process more difficult feed (higher N, lower API) while maintaining yield throughout the cycle
- Diesel T95 was 10 °F higher along with higher consumption and thus increased volume swell

AIChE
Southwest Process Technology Conference

8th AIChE Southwest Process Technology Conference I October 6-7, 2016 I Galveston, TX

Resolve dP growth with optimized grading

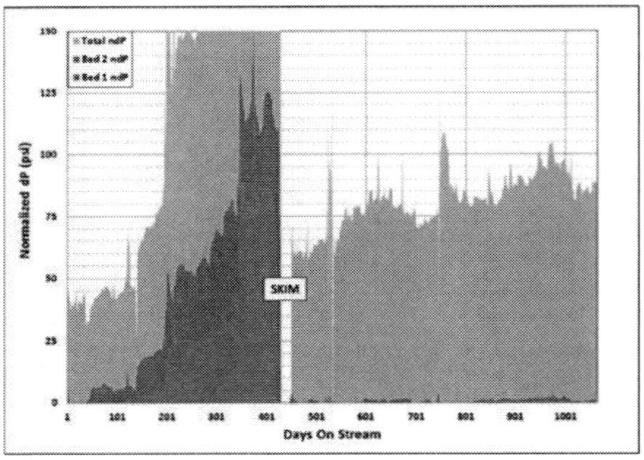

- This unit suffered from chronic delta P growth
- Top two beds were skimmed halfway through the cycle due to high dP
- Delta P increase has been greatly reduced since the skim as a result of grading scheme optimization (smaller catalyst in filter tray, increased grading volume and layering)
- Unit is on track to achieve 60 months cycle life compared to original target of 36 months

AIChE Southwest Process Technology Conference

Summary

- In addition to catalyst activity, stability is a key performance indicator to ensure units meet turnaround timing and minimize unplanned downtime
- New catalysts enable refiners to process higher value and more difficult feed streams without sacrificing cycle life
- Units constrained by metals poisoning or delta P growth can be improved by a robust demet system or an optimized grading scheme and hardware options such as Shell Global Solutions HDTrays™ & Filter Trays™
- These products combined with Criterion's technical expertise help refiners maximize value from their assets

AIChE Southwest Process Technology Conference

MEETING ADDITIONAL PLANT CAPACITY REQUIREMENTS WITHOUT ADDING ADDITIONAL VESSELS

Gene Eberhardt
Vice President – Reactor Division
Woven Metal Products

Agenda

- Case Study
- Standard Equipment CFD Review
- Outlet Support Grid CFD Review
- Outlet Support Grid Design Overview
- Outlet Support Grid Benefits
- Q & A

OUTLET SUPPORT GRID (OSG)

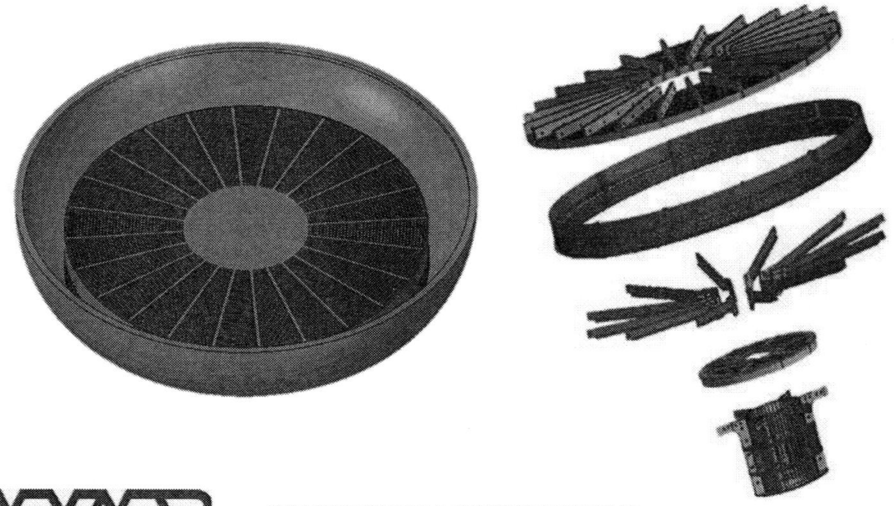

WOVEN METAL PRODUCTS, INC

CONFIDENTIAL PRESENTATION

US Patent Protected

AIChE Southwest Process Technology Conference

8th AIChE Southwest Process Technology Conference I October 6-7, 2016 I Galveston, TX

CASE STUDY

- Project Located In West Virginia
- Three Mole Sieve Bed System
- Standard Outlet Baskets Installed
- Previous Containment Issues
- Poor Bed Utilization
- Pre-Installation Bed 130 psi
- Plant Conditions Warrant ESD

WOVEN METAL PRODUCTS, INC

AIChE Southwest Process Technology Conference

8th AIChE Southwest Process Technology Conference I October 6-7, 2016 I Galveston, TX

CASE STUDY

- Project Success In Areas Of:
  - Safety
  - Project Execution
  - Process Improvement
  - ROI
- Three Outlet Support Grids Installed Within 1.5 Hours For Each Bed – Far Exceeding Maintenance Plan

WOVEN METAL PRODUCTS, INC

AIChE
Southwest Process Technology Conference

8th AIChE Southwest Process Technology Conference I October 6-7, 2016 I Galveston, TX

CASE STUDY

- Post-Installation Bed Reduced To 3 psi
- Significant Reduction In Support Media
- Additional Mole Sieve Volume
- Containment Issues Resolved
- Project Payback Estimated Less Than 12 Months
- Outstanding Project ROI

WOVEN METAL PRODUCTS, INC

AIChE
Southwest Process Technology Conference

8th AIChE Southwest Process Technology Conference I October 6-7, 2016 I Galveston, TX

CFD STUDY

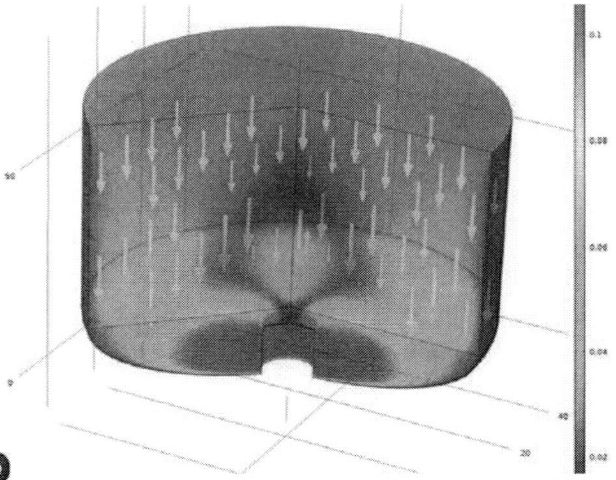

WOVEN METAL PRODUCTS, INC

AIChE Southwest Process Technology Conference

8th AIChE Southwest Process Technology Conference I October 6-7, 2016 I Galveston, TX

CFD STUDY

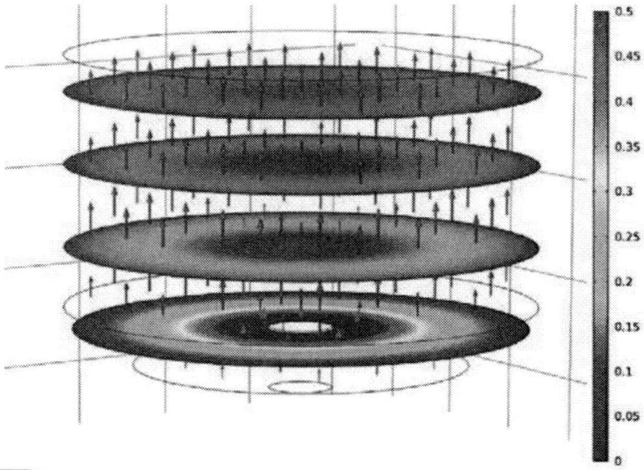

WOVEN METAL PRODUCTS, INC

AIChE Southwest Process Technology Conference

8th AIChE Southwest Process Technology Conference I October 6-7, 2016 I Galveston, TX

CFD STUDY

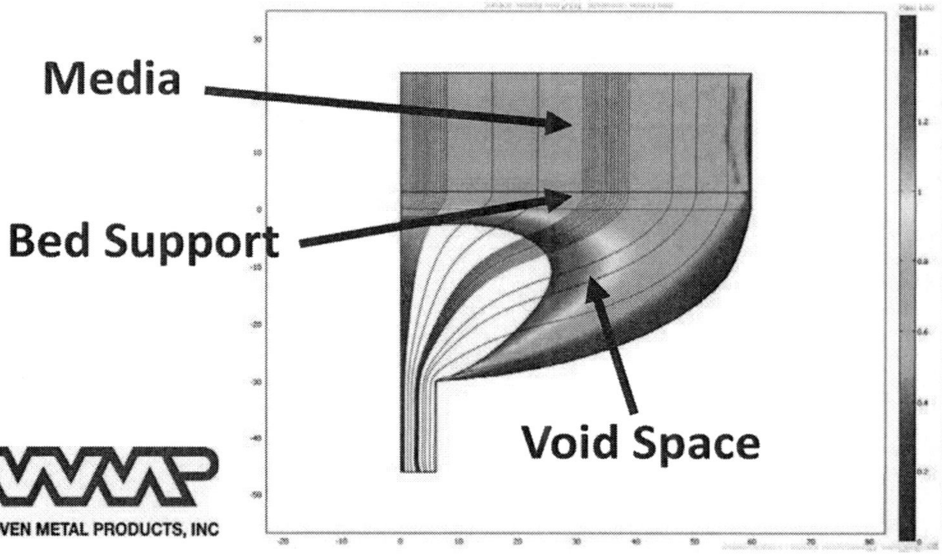

WOVEN METAL PRODUCTS, INC

8th AIChE Southwest Process Technology Conference I October 6-7, 2016 I Galveston, TX

CFD STUDY

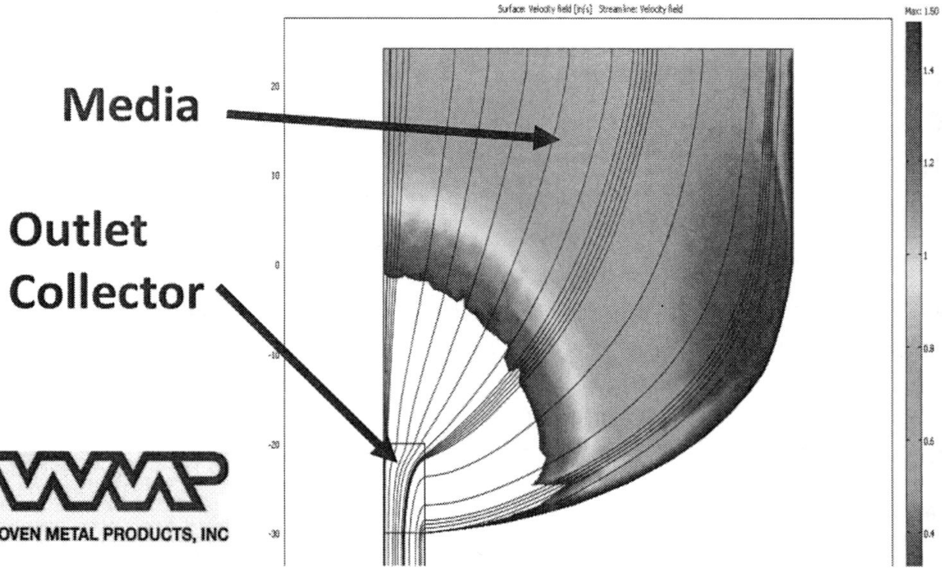

WOVEN METAL PRODUCTS, INC

8th AIChE Southwest Process Technology Conference I October 6-7, 2016 I Galveston, TX

CFD STUDY

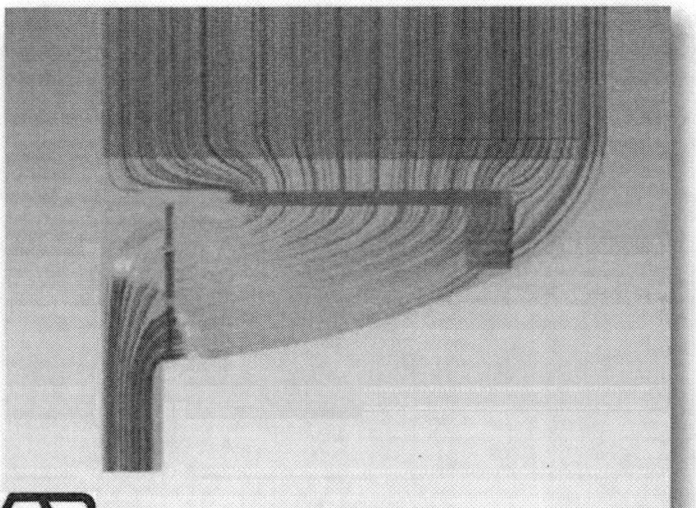

WOVEN METAL PRODUCTS, INC

AIChE Southwest Process Technology Conference

8th AIChE Southwest Process Technology Conference I October 6-7, 2016 I Galveston, TX

OUTLET SUPPORT GRID (OSG)

- All New Equipment
  - Elephant Stool
  - Outlet Connector Rods
  - Outer Skirt Sections
  - Top Support Grid Screen
  - Hardware

WOVEN METAL PRODUCTS, INC

AIChE Southwest Process Technology Conference

8th AIChE Southwest Process Technology Conference I October 6-7, 2016 I Galveston, TX

OUTLET SUPPORT GRID (OSG)

- No Rope Packing Required
- Performs Well Over Thermal Cycles
- No Support Media Or Catalyst Migration Issues

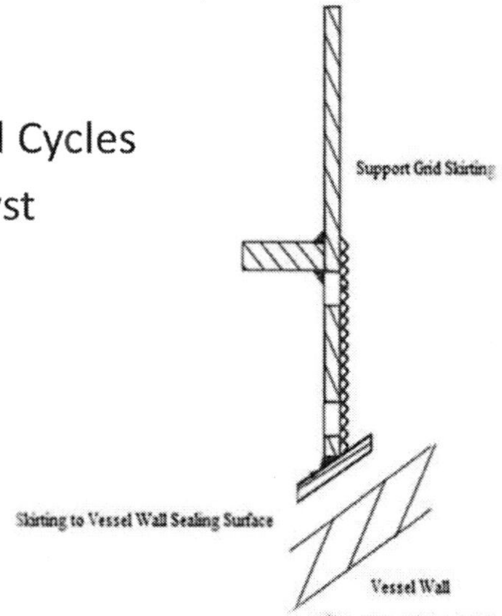

WOVEN METAL PRODUCTS, INC

AIChE
Southwest Process Technology Conference

8th AIChE Southwest Process Technology Conference I October 6-7, 2016 I Galveston, TX

INSTALLATION

- All Pieces Manway Installable
- No Welding Required
- Minimal Hardware
- Same Size Bolting Throughout Design
- Minimum Installation Time
- Easily Retrofitted
- Horizontal Limiter Lugs Mitigate Horizontal Movement

WOVEN METAL PRODUCTS, INC

AIChE
Southwest Process Technology Conference

8th AIChE Southwest Process Technology Conference I October 6-7, 2016 I Galveston, TX

INSTALLATION

WOVEN METAL PRODUCTS, INC

AIChE

Southwest Process Technology Conference

8th AIChE Southwest Process Technology Conference I October 6-7, 2016 I Galveston, TX

COST SAVINGS

- Maximum Bed Utilization
- Capture Void Space From Standard Bed Support
- Increase Bed Volume
- Full Bed Containment
- Reduced Pressure Drop

WOVEN METAL PRODUCTS, INC

AIChE

Southwest Process Technology Conference

8th AIChE Southwest Process Technology Conference I October 6-7, 2016 I Galveston, TX

COST SAVINGS

- Reduced Regenerations
- Longer Run
- Ease Of Maintenance
- Less Valve Cycles
- 50 – 75 % Reduction Of Required Support Media

WOVEN METAL PRODUCTS, INC

AIChE Southwest Process Technology Conference

8th AIChE Southwest Process Technology Conference I October 6-7, 2016 I Galveston, TX

INSPECTION ACCESS

- Manway Hatch Allows Access To Outlet Piping
- Removable “Pie” Sections Eliminates Need To Remove Entire Support Grid For Inspections

WOVEN METAL PRODUCTS, INC

AIChE Southwest Process Technology Conference

8th AIChE Southwest Process Technology Conference I October 6-7, 2016 I Galveston, TX

WOVEN METAL PRODUCTS, INC

AIChE
Southwest Process Technology Conference

8th AIChE Southwest Process Technology Conference I October 6-7, 2016 I Galveston, TX

Q & A

WOVEN METAL PRODUCTS, INC

AIChE
Southwest Process Technology Conference

8th AIChE Southwest Process Technology Conference I October 6-7, 2016 I Galveston, TX

THANK YOU!

GENE EBERHARDT

Vice President

Reactor Division

281 – 331 – 4466 Office

281 – 682 – 8058 Mobile

GCE@wovenmetal.com

WOVEN METAL PRODUCTS, INC

AIChE Southwest Process Technology Conference

8th AIChE Southwest Process Technology Conference I October 6-7, 2016 I Galveston, TX

Woven Metal Products

WOVEN METAL PRODUCTS, INC

1201 FM 517
Alvin, TX 77511

AIChE Southwest Process Technology Conference

8th AIChE Southwest Process Technology Conference I October 6-7, 2016 I Galveston, TX

Catacel

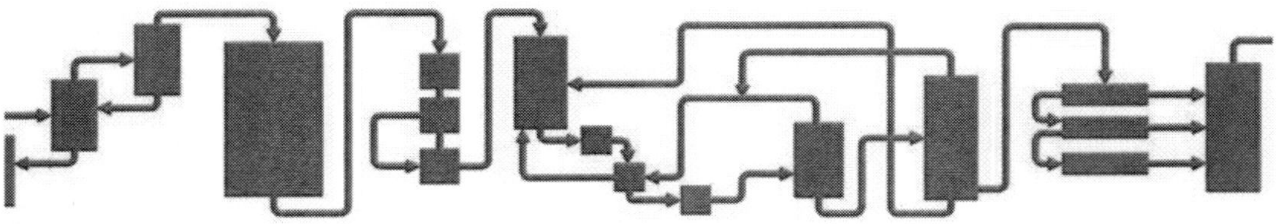

Redefining Steam Methane Reformer Performance in Hydrogen Production

Diane Dierking
Johnson Matthey Process Technologies
Oakbrook Terrace, Illinois

JM Johnson Matthey Process Technologies

Heat Transfer – Tubular Reactors

JM Johnson Matthey Process Technologies

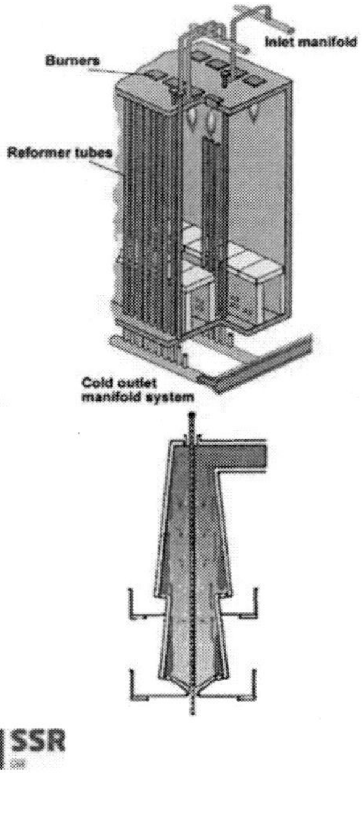

Catacel SSR

What is CATACEL$_{JM}$ Technology?

Johnson Matthey
Process Technologies

Catalysts & absorbents

Foils

Catacel SSR

CATACEL$_{JM}$ SSR Technology
Controlled Gas Flow and Wall Impingement

Johnson Matthey
Process Technologies

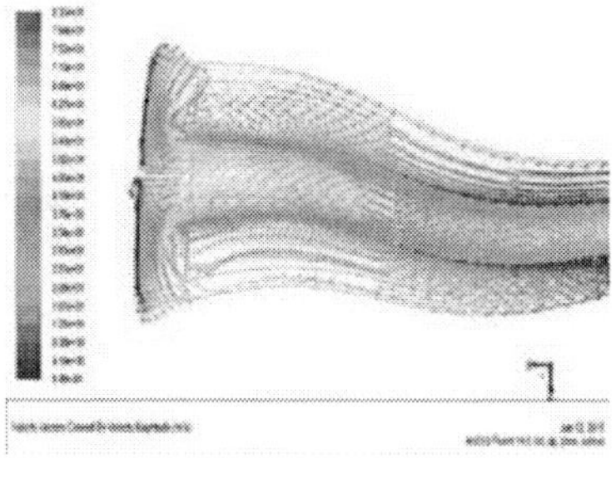

Catacel SSR

Novel Technology

JM Johnson Matthey Process Technologies

- **CATACEL$_{JM}$ SSR** is a novel, patent-protected steam methane reforming technology that provides leading-edge performance
- This technology delivers step-out increases in heat transfer, pressure drop and activity, which give a range of plant-specific operational benefits

Catacel SSR

CATACEL$_{JM}$ SSR Technology Enhanced Heat Transfer

JM Johnson Matthey Process Technologies

- Heat transfer, traditional pellets
  - Limited by static gas film at tube wall
  - Smaller pellets will help break up the gas film, but increase dP

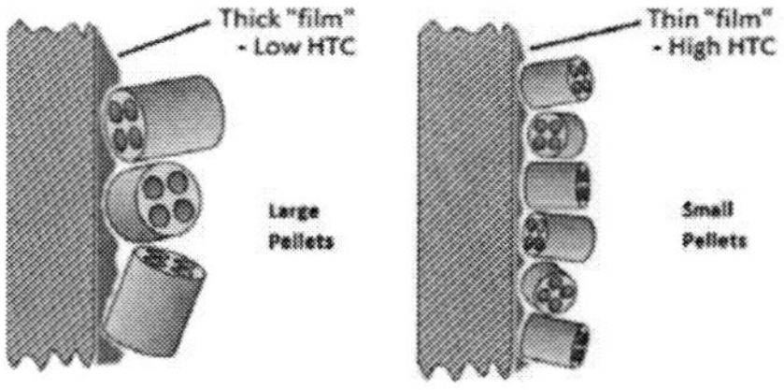

- Highest heat transfer ensures heat is available where it is needed to drive reactions
  - 20% to 30% increase in heat transfer

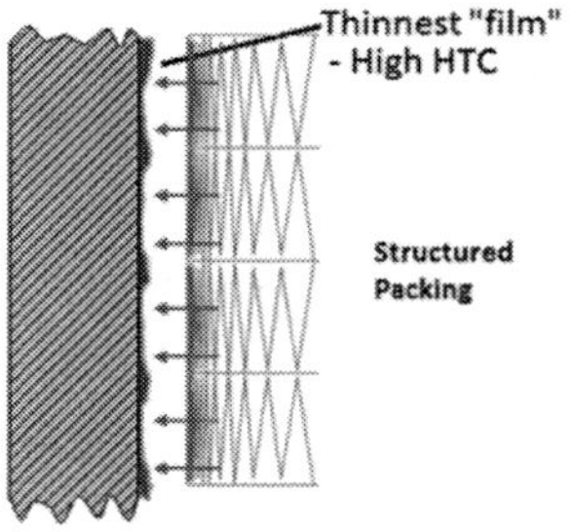

Catacel SSR

CATACEL$_{JM}$ SSR Technology
Step-Change in Performance Characteristics

Johnson Matthey
Process Technologies

- Higher geometric surface area
  - 40% more surface area means more activity
  - Resilient to upsets
  - Higher conversion or more throughput at same fuel rate
- Increased void space with **CATACEL$_{JM}$ SSR**
  - Pressure drop 10% to 20% lower
  - Lowers cost of compression and higher throughput for same pressure drop

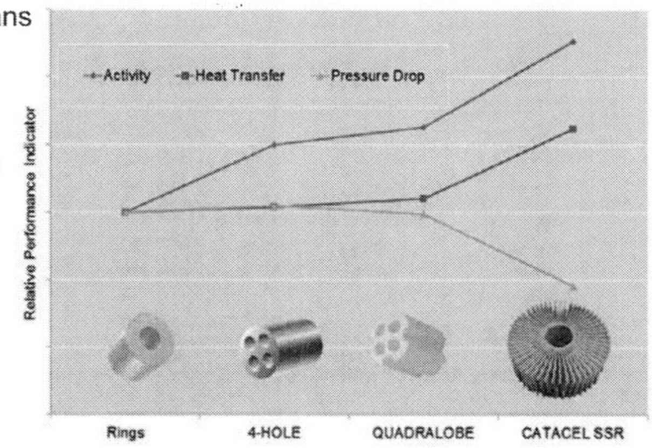

Catacel SSR

CATACEL$_{JM}$ SSR Technology
Different Than Pellets

Johnson Matthey
Process Technologies

- **CATACEL$_{JM}$ SSR** design is tuneable to generate maximum value
  - Fan dimensions provide a number of degrees of freedom
  - Easy to adjust design, enhancing desired key performance characteristics
    - Heat Transfer
    - Geometric Surface Area (Activity)
    - Voidage (Pressure drop)

Catacel SSR

Revamp – Increased Throughput

Johnson Matthey
Process Technologies

- Hydrogen plant switched from naphtha to natural gas feed
- Operation on natural gas has moved the plant to a TWT limit
- 15% more H_2 now required
- **CATACEL$_{JM}$ SSR** heat transfer and activity deployed to minimize TWT
- Lower dP of **CATACEL$_{JM}$ SSR** allows operation at 115% rate with same pressure drop as pellets at 100% rate

Case	Units	Base	1	2
Catalyst		Pellets	CATACEL$_{JM}$ SSR	CATACEL$_{JM}$ SSR
Description		Base	Same feed Lower firing	More feed Extra firing
Relative feed flow	%	100.0	100.0	115.0
Relative fuel flow	%	100.0	99.5	112.5
Relative comb'n air flow	%	100.0	99.4	105.0
Excess air	%	15.0	15.0	7.5
Exit temperature	°F	1580	1580	1582
Methane slip	mol%dry	2.80	2.75	2.80
Max TWT	°F	1634	1625	1636
Pressure drop	psi	18.7	14.6	18.6
Fluegas exit temperature	°F	1827	1816	1854
Relative hydrogen make	%	100.0	100.2	115.0

Catacel SSR

Benefit Summary – Revamp/Throughput

Johnson Matthey
Process Technologies

- **CATACEL$_{JM}$ SSR** is a novel, patent-protected steam methane reforming technology that provides leading-edge performance
- This technology delivers step-out increases in heat transfer, pressure drop and activity, which give a range of plant-specific operational benefits

- **For revamps and plant uprates:**
  - ✓ 10% - 20% throughput increase with no change in tube wall temperature margin
  - ✓ Tubes can be replaced with those having thinner walls (lower cost per tube)

Catacel SSR

Operating Units – Fuel Savings

Johnson Matthey
Process Technologies

- **$CATACEL_{JM}$ SSR** deployed as direct replacement for pellets
- Same feed rate and unchanged plant operating conditions
- Enhanced heat transfer performance of **$CATACEL_{JM}$ SSR** allows less fuel to be consumed to achieve the same TWT
- Savings taken via a reduction in the natural gas fuel rate
- Considerable savings – maximized in high gas cost regions

	Units	Top Fired Reformer	
Catalyst Type		Pellets	$CATACEL_{JM}$ SSR
Nat Gas price $/MBtu		9.48	9.48
Hydrogen Product	MMSCFD	127.2	127.2
Nat Gas Feed Rate	MSCFH	2,071	2,071
Nat Gas Fuel to Burners	MSCFH	88.0	81.1
Off-Gas to Burners	MSCFH	2,602	2,602
Catalyst volume	ft³	1990	1990
Nat Gas Feed	$M/year	178.0	178.0
Nat Gas Fuel	$M/year	7.6	6.9
Saving on Nat Gas Fuel	$/yr		$ 697,001
	%		7.9

Catacel SSR

Reference – Increased Flow Rate

Johnson Matthey
Process Technologies

- 20-25% increase in achievable production rates
  - Highest sustained production rate with pellets was 71,000 SCFH (1,902 Nm^3/hr)
  - Maximum rate with **$CATACEL_{JM}$ SSR** is 90,000 SCFH (2,411 Nm^3/hr)
    - Several campaigns > 85,000 SCFH (2,277 Nm^3/hr)

H_2 Production Rate (scfh)	Pellet (% of Operating Time)	$CATACEL_{JM}$ SSR (% of Operating Time
40,000 – 50,000	24%	8%
50,000 – 60,000	31%	26%
60,000 – 70,000	34%	27%
70,000 – 80,000	N/A	21%
80,000 – 90,000	N/A	11%

Catacel SSR

Reference – Fuel Savings

Johnson Matthey
Process Technologies

- Fuel savings of 57%
  - Pellet: 6.6 SCFH H_2 / SCFH NG consumed as fuel
  - SSR: 15.5 SCFH H_2 / SCFH NG consumed as fuel

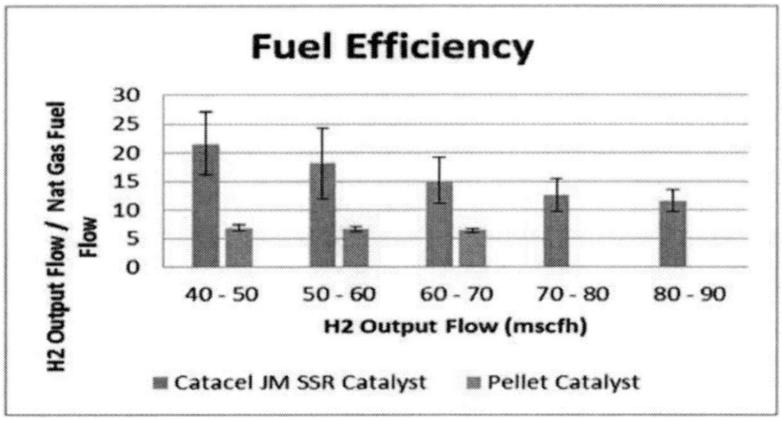

Catacel SSR

Operating Units – Remove Process Constraints

Johnson Matthey
Process Technologies

- **CATACEL$_{JM}$ SSR** deployed as direct replacement for pellets
- Same feed rate and unchanged plant operating conditions
- Enhanced heat transfer performance of **CATACEL$_{JM}$ SSR**
  - Less fuel consumed
  - Lower TWT and higher TWT margins
- Enhanced activity of **CATACEL$_{JM}$ SSR**
  - Operation outside carbon forming zone without the use of alkalized catalyst

	Units	57-4Q: 57-4XQ	CATACEL$_{JM}$ SSR
Feed Flow Rate	MSCFH	315.4	315.4
Steam to Carbon Ratio	mol/mol	2.8	2.8
Inlet Temperature	°F	842	842
Off-Gas	MSCFH	157.7	157.7
Nat Gas Fuel	MSCFH	158.5	156.4
	%		1.3
Exit Temperature	°F	1592	1589
Exit Pressure	psig	346.7	346.7
Catalyst Pressure Drop	psi	17.7	20.7
Flue Gas Temperature	°F	1813	1795
Max Tube Wall Temperature	°F	1637	1623
	Δ °F		-14
Peak Heat Flux	Btu/hr/ft²	42326	43153
Approach to Equilibrium	°F	4.7	2.0
Minimum Carbon Margin	°F	-38	37

Catacel SSR

Operating Units – Steam:Carbon Ratio

JM Johnson Matthey
Process Technologies

- **CATACEL$_{JM}$ SSR** deployed as direct replacement for pellets
- Equivalent hydrogen production
- Enhanced activity of **CATACEL$_{JM}$ SSR**
  - Maintains comfortable carbon margin
- Enhanced heat transfer properties of **CATACEL$_{JM}$ SSR**
  - Impact of additional firing on TWT is minimized

	Units	QUADRALOBE	CATACEL$_{JM}$ SSR	
Steam to Carbon Ratio	mol/mol	2.9	2.9	2.7
Feed Flow Rate	%	Base	99.9	102.4
Fuel Flow Rate	%	Base	98.9	91.5
Total NG Flow Rate	%	Base	99.6	99.4
Exit Temperature	°F	1521	1521	1521
Max Tube Wall Temperature	°F	1593	1582	1582
Carbon Margin	°F	Base	+36	+23
Catalyst Pressure Drop	psi	38.4	38.4	37.0
Process Heat Duty	%	Base	99.8	96.2
Hydrogen Production	%	Base	Base	Base

Catacel SSR

Benefit Summary – Existing Plants

JM Johnson Matthey
Process Technologies

- **CATACEL$_{JM}$ SSR** is a novel, patent-protected steam methane reforming technology that provides leading-edge performance
- This technology delivers step-out increases in heat transfer, pressure drop and activity, which give a range of plant-specific operational benefits

- **For existing plants**:
  - ✓ 1% - 2% throughput increases (at same trim fuel rate) when **CATACEL$_{JM}$ SSR** used as direct replacement for standard pellets
  - ✓ 2% - 10% trim fuel savings when **CATACEL$_{JM}$ SSR** used as direct replacement for standard pellets
  - ✓ 20% decrease in pressure drop compared to standard pellets
  - ✓ Steam to Carbon decrease of 0.5 units (mol/mol) when replacing standard non-alkalized pellets

Catacel SSR

Case Study – New Hydrogen Plant

Johnson Matthey
Process Technologies

Top-Fired	Units	Pellets	CATACEL$_{JM}$ SSR
Number tubes	–	460	385
Tube ID	in	4.50	4.61
Tube heated length	ft	41.8	46.4
Furnace length	ft	48.0	50.0
Furnace width	ft	76.5	61.5
Pressure drop	psig	29.9	30.4
Peak TWT	–	Base	Base

Catacel SSR

Benefit Summary – New Plants

Johnson Matthey
Process Technologies

- **CATACEL$_{JM}$ SSR** is a novel, patent-protected steam methane reforming technology that provides leading-edge performance
- This technology delivers step-out increases in heat transfer, pressure drop and activity, which give a range of plant-specific operational benefits

- **For new build projects**:
  - ✓ Decrease tube count by 10% - 20%
  - ✓ Decrease tube wall thickness, longer tubes
  - ✓ 5% - 20% lower capital cost of the radiant box

Catacel SSR

Conclusions

Johnson Matthey
Process Technologies

- Novel steam reforming technology, **CATACEL$_{JM}$ SSR**
  - Provides improvement in heat transfer (overcomes limitations), activity, and pressure drop
  - Carbon-free operation over wider range of operating conditions
  - Tolerates upset conditions that would normally lead to unscheduled change-out, whether low S:C or poisoning
  - Allows increase in throughput up to limitations of burners and convection section
  - OPEX savings on fuel
  - Unique properties allow for CAPEX reduction in new plant designs through optimized furnace designs

Catacel SSR

Thank you!

JM
Johnson Matthey
Process Technologies

Diane Dierking
2 Transam Plaza Drive
Suite 230
Oakbrook Terrace, IL 60181
(630) 268-6300

Catacel SSR

CD-adapco
A Siemens Business

Coupled Simulations in Steam Methane Reformers using CFD and Reaction Kinetics

Niveditha Krishnamoorthy (Combustion Tech-Specialist)
Chandraprakash Tourani (Sr. Application Engineer)
Ravindra Aglave, (Director - Chemical & Process Industry)

Steam Methane Reformer

CD-adapco
A Siemens Business

- Steam Reforming: A method for producing hydrogen , carbon monoxide from hydrocarbon fuels like natural gas
- The hydrogen produced can be used for:
  - Feedstock for fuel cells
  - Hydrogenating vegetable oils for food industry
  - Industrial synthesis of ammonia (Haber process)
  - Hydrocracking of heavy petroleum fractions into lighter ones
  - Hydro-desulfurization for sulfur removal from natural gas and other refined petroleum products

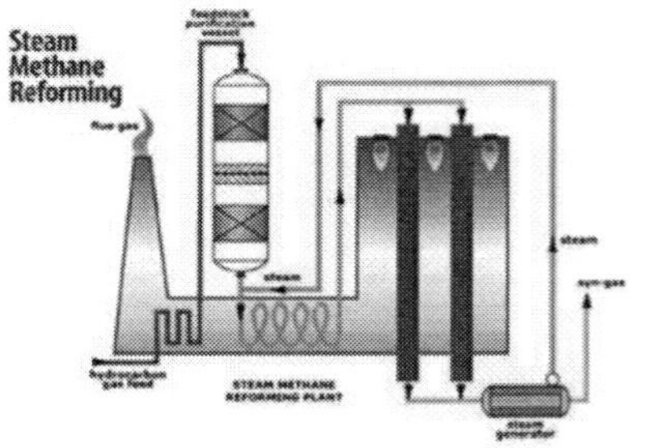

http://www.eajv.ca/english/h2

High temperature (700-1100 deg C)

$CH_4 + H_2O \rightleftharpoons CO + 3\,H_2$ [Endothermic]

(in the presence of a metal based catalyst)

Low Temperature water gas shift reaction

$CO + H_2O \rightleftharpoons CO_2 + H_2$ [Mildly Exothermic]

Modeling Challenges

- Typical reformer furnace could have large number of burners
- Several hundred process tubes present
- Furnace dimensions are orders of magnitude larger than process tube dimensions
- Burners, process tube spacing influences flow patterns and heat flux distribution.
  - Hydrogen conversion rates affected
  - Tube wall temperatures that could lead tube failures if too high

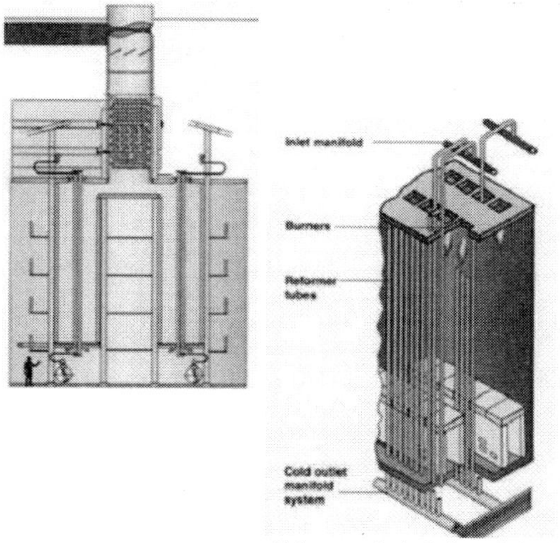

Hydrogen production by Steam Reforming,
Ray Elshout, Chemical Engineering, May 2010

Modeling Methodologies

- Current codes other than STAR-CCM+
  - Highly simplified representation of firebox and 1D modeling of process side
    - Unable to capture geometry related influences
      - Recirculation zones inside the furnace - Flue gas mal distribution
      - Shadowing effects on process tubes
      - Hot spots on process tubes leading to tube failures
- STAR-CCM+
  - 3D modeling of burner side [Firebox] and 3D modeling of process side
    - Computationally expensive but possible
  - 3D modeling of Firebox and 1D modeling of process side (Reacting Channel Co-simulation)
    - Elegant way of coupling firebox side physics to tube side physics in a computationally efficient manner
    - Geometry related influences on tube wall temperatures and conversion rates effectively captured

Reacting Channel Co-Simulation in STAR-CCM+

- Gas-Phase:
  [FireBox Side]
  - 3-D, turbulent flow
  - Combustion modeling
  - Heat transfer modeling
    - Conduction
    - Convection
    - Radiation
- Reacting Channel:
  [Process Side]
  - 1-D Plug Flow Reactor (PFR)
  - Process-side kinetics modeling
  - Packed bed physics for accurate heat transfer and pressure drop calculations

Model Furnace represented in 3-D

Process side physics in 1-D

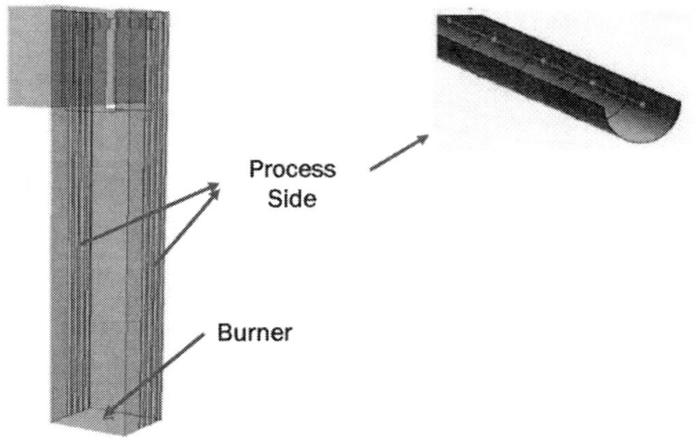

Furnace Side Modeling

- 3-D CFD Calculations to get:
- Temperature / Heat distribution in furnace box
- Tube skin temperature
- Flame shape and length
- FGR pattern
- Radiant : Convective section heat balance
- Emissions
  - NOx
  - CO

3-D and 1-D coupling

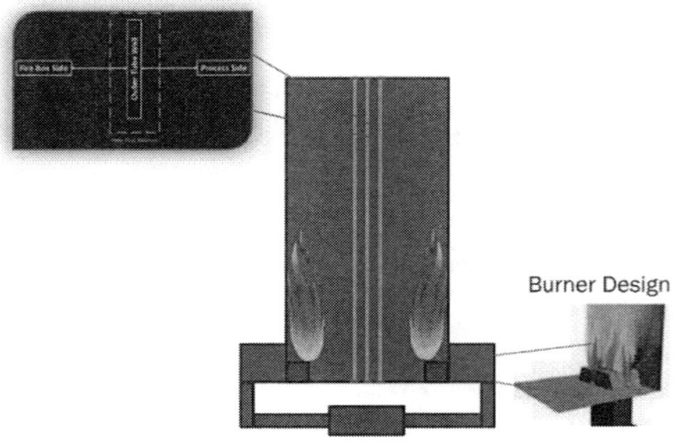

Firebox Side Results

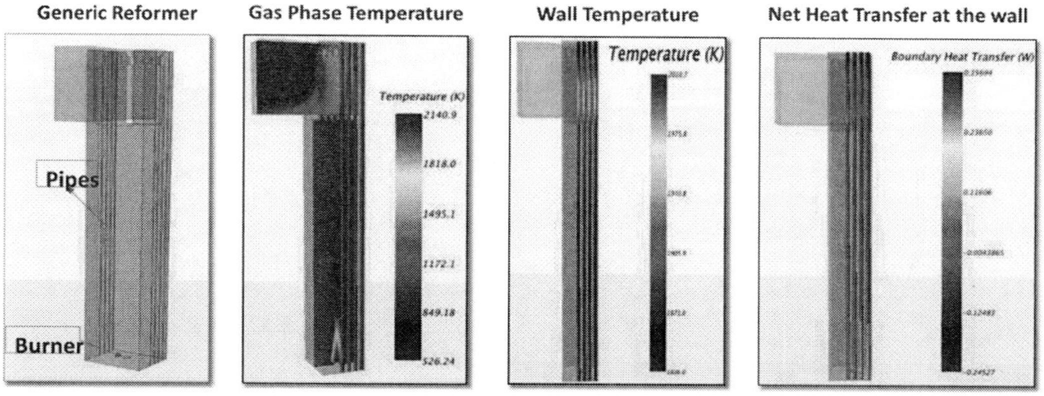

Tube Side Physics –(1)

- Heat Transfer Coefficient Computation
  - Simple pipe
  - Packed beds
    - Leva/Grummer
    - Beek
    - DeWasch/Froment
  - User-defined tabular input
- Heat transfer through a packed bed has a significant effect on the performance of the equipment
- Much higher heat transfer coefficient values seen in packed beds than simple pipes. This significantly influences conversion rates

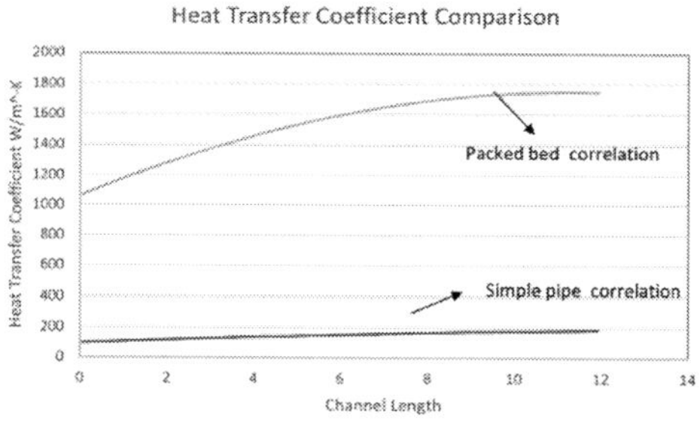

Tube Side Physics –(2)

CD-adapco
A Siemens Business

- Pipe friction correlations for packed beds and simple pipes available
- Accurate pressure drop through packed tubes can be captured

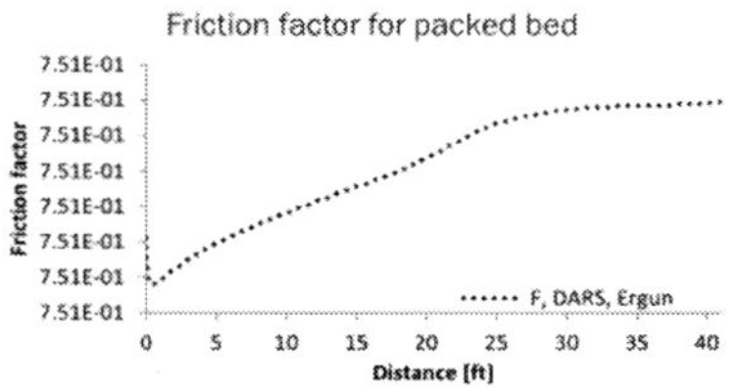

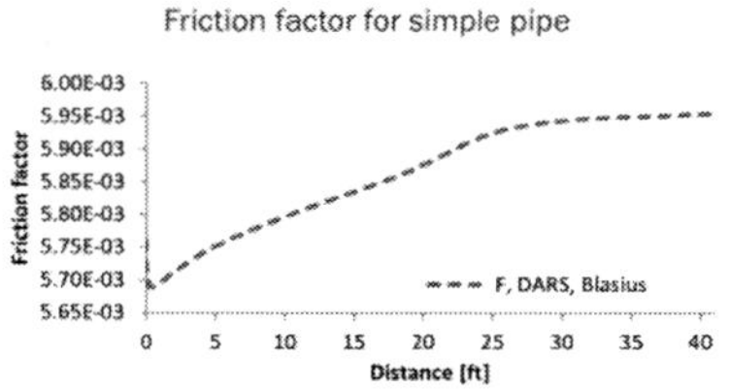

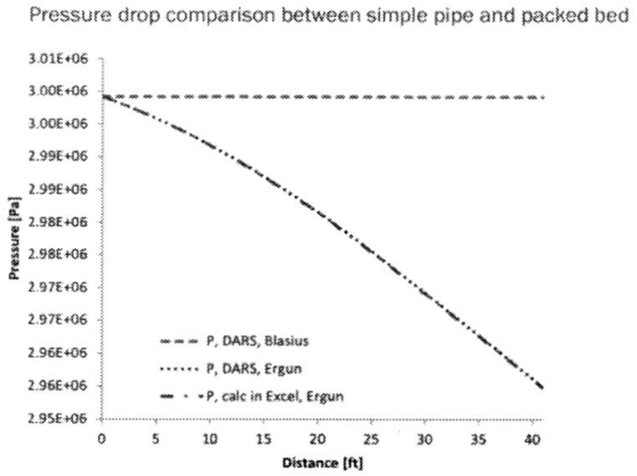

Tube Side Physics –(3)

CD-adapco
A Siemens Business

- Steam-Methane reforming kinetics
  - Detailed chemistry
  - Reduced chemistry
  - User-defined kinetic rates
- Output Quantities
  - Methane conversion
  - Hydrogen yield
  - Process fluid temperature
  - Tube wall temperature

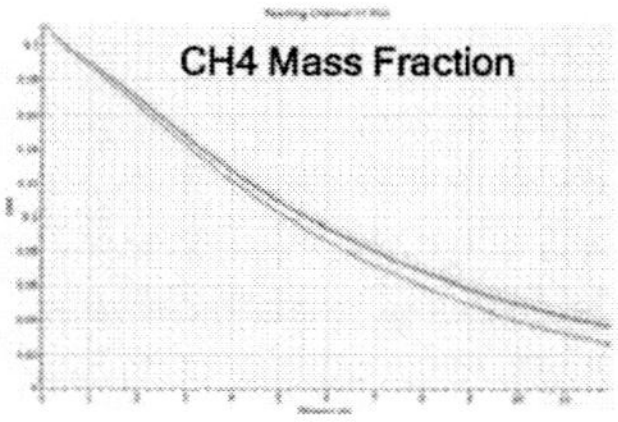

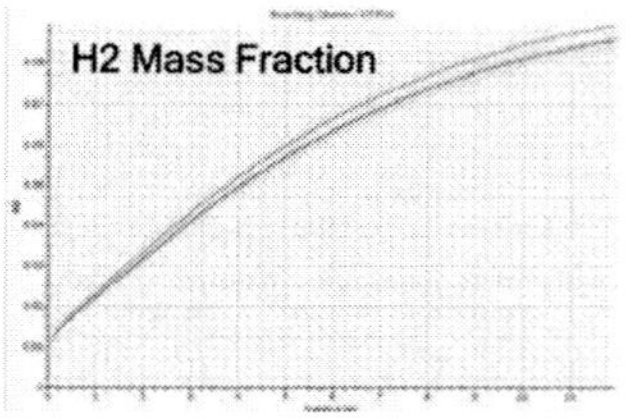

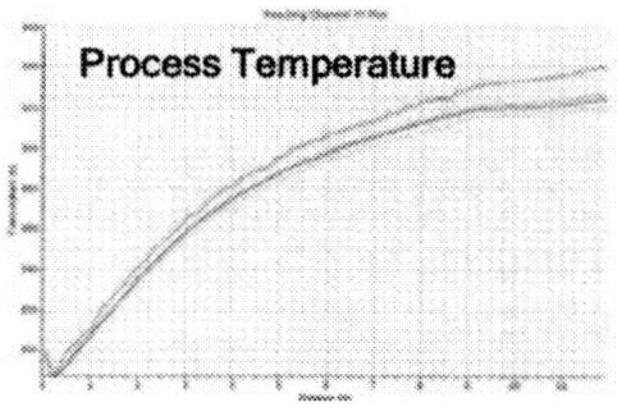

Coupling between Firebox and Process Side

Coupling achieved through energy balance at the outer tube walls

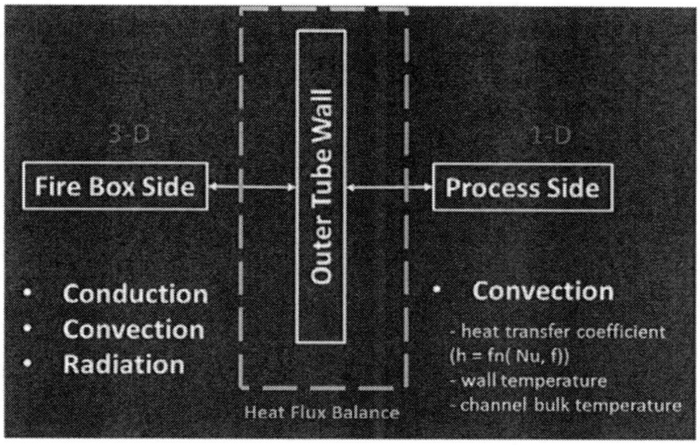

Summary

- An adequate representation of the physics on the firebox side is possible using STAR-CCM+
- Correlations for heat transfer coefficient and pressure drop enables simulations of catalytic processes like SMR to be effectively simulated using reacting channel co-simulation
- Flexible kinetics description for process side enables users to optimize process side chemistry for accurate description of reactant conversion and product yields

Gasoline Blending Direct to Ship

Seiver, D.,

Originally presented on: 10/6/2016 9:30:00 - 9:55:00

A new control system and single-blend optimization system make it possible to blend gasoline directly into tanker ships for delivery, rather than only into tanks, thereby saving millions in capital and reducing product giveaway. In addition to terminal equipment, the facility has eight berths for ships and more than 100 tanks of various sizes with total capacity of 10.5 MMbbl. The complex can facilitate 10 crude ships per month. This presentation will highlight how automation overcame several challenges with gasoline blending direct to ship.

Capturing Operational Value Through Data Analytics

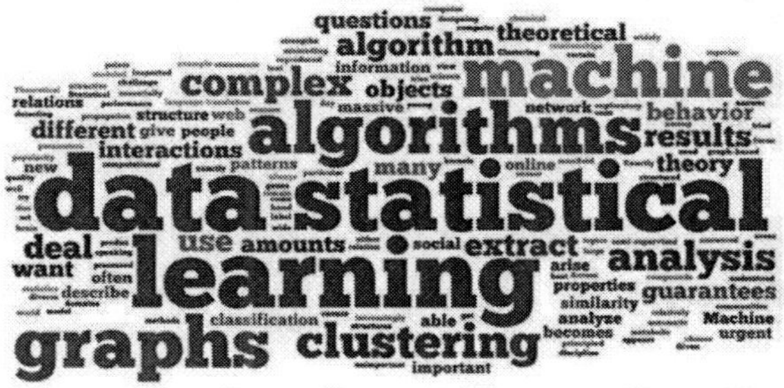

Doug White
Principal Consultant
Emerson Process Management
Houston, Texas

1

Southwest Process Technology Conference

Speaker

Doug White (doug.white@emerson.com)

Principal Consultant

PlantWeb Solutions Group

Emerson Process Management

Background: Many years experience designing, justifying, installing and commissioning advanced real time computer/ automation applications in the process industries.

EMERSON.
Process Management

AIChE
Southwest Process Technology Conference

2

Data is Increasing

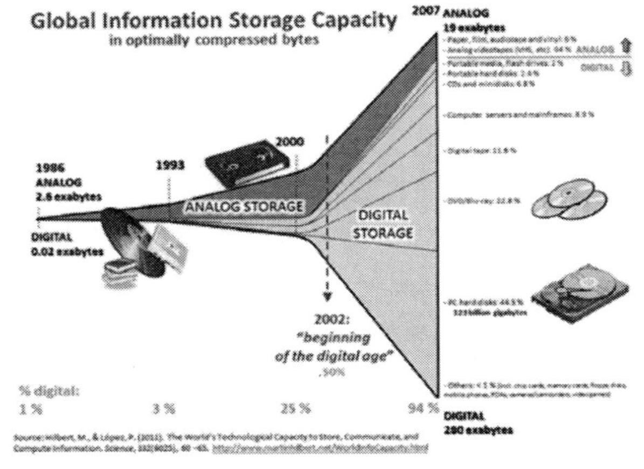

Exponential growth of global storage capacity
and processing power

Data Is Increasing

- Walmart's Online Database
  - More than 250 billion rows
  - More than 20 million updates per hour
  - 3 years history available online
  - 24,000 queries per hour from 1500 users
- Internet Devices
  - Gardner group estimates 5 billion connected devices in 2015 – 25 billion by 2020
  - Plant device connectivity is increasing similarly
- Process Company
  - 10 million tags across 15 sites
  - Implementing system to provide real time access to 3 years of 1 minute sampled historical data for tags
- Refining Company
  - Reported total of 80 billion data items from four sites in one year

Types of Data Analytics

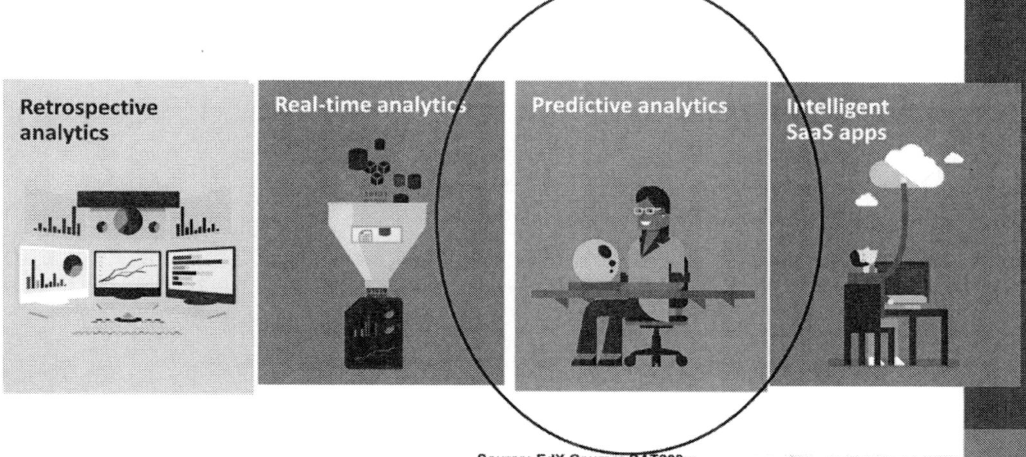

Source: EdX Course: DAT203x: Data Science and Machine Learning Essentials, Ch.2; Microsoft

5

Creating Value From Data - Predictive Analytics – Looking Forward

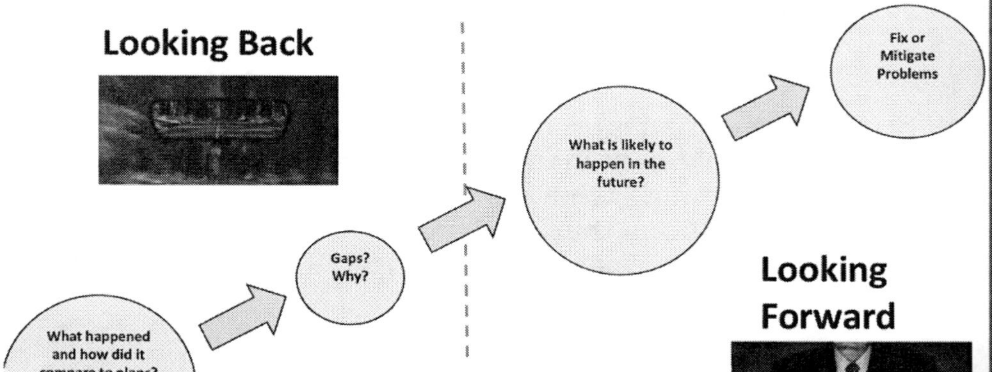

Predictive analytics, in general, is the use of current and historical data in conjunction with system models to predict future trends in real-time (near).

6

Where is Predictive Analytics in the “Hype” Evolution?

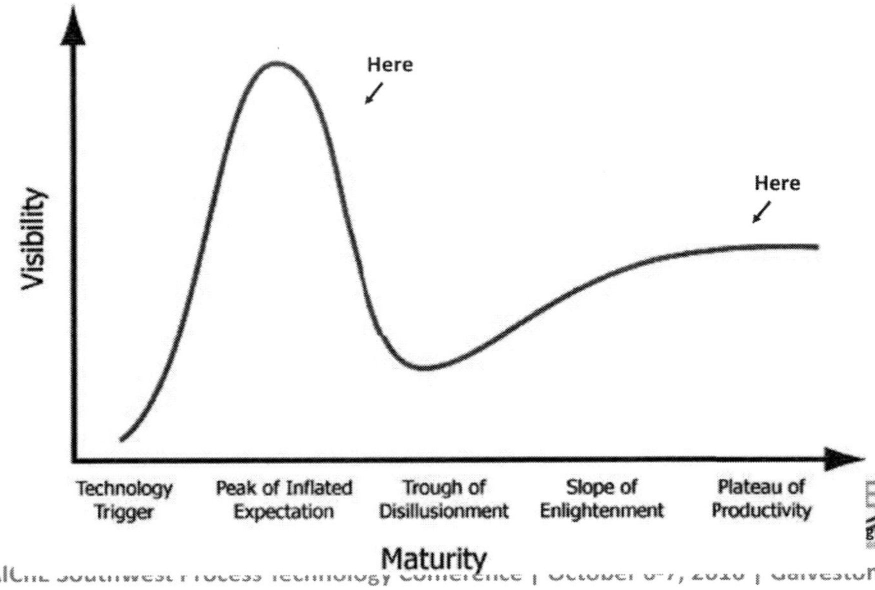

Predictive Analytics - Moneyball

- In 2001, the Oakland A’s (salaries $44m) lost to the New York Yankees (salaries $125m) in the divisional playoffs and then lost 3 top players to free agency.
- Billy Beane, the manager, hires Peter Brand, a new graduate, to implement statistical analysis of previous seasons to determine what it takes to win. He develops correlations that go against some of the traditional baseball ideas. Beane supports using the approach to pick new players.
- In 2002, The A’s set a league record by winning 20 games in a row on the way to winning 95 games overall and making the league playoffs – and doing this again in 2003.

Predictive Analytics - Predicting Supreme Court Decisions

- Andrew Martin, a professor of political science at Washington University, developed a statistical model of supreme court decisions based on actual decisions from 1994 to 2001 (no change in justices during that period.
- He used the model to predict decisions for the 2002 term. Results were compared against a panel of distinguished legal individuals experienced with the supreme court.
- For the 68 cases in the 2002 term:
  - Model accuracy – 75%
  - Experts accuracy – 59%

AIChE Southwest Process Technology Conference

AFPM American Fuel & Petrochemical Manufacturers

How Does This Apply (And Not Apply) To The Process Industries?

Plant ("Near") Real Time Data Includes Many Types

Numeric Data	Spectral Data	Video/ Images	Text
Process/ Offsites/ Utilities/ Terminal Data	Analyzers – NIR, etc	Cameras – Flares, Process, Security, Etc	Daily Plans, Notes, etc
Lab Data	Rotating Machinery – Vibration, etc.	Thermography	Maintenance - Inspection, Work Orders, etc
Accounting, KPI's, Planning, Scheduling, etc	Infrared/ Acoustic/ Ultrasonic – Event Detection	Misc	E-Mail

Typical Refinery DCS IO (Input / Output) Numbers – Continued Increase

1990 – "Hard" IO Count: 20,000

2015 – "Hard" IO Count: 50,000

2025 – "Hard" IO Count: 100,000+

Timeline

Photo Source: US DOE

Remote Access - "Internet of Things" in Plants

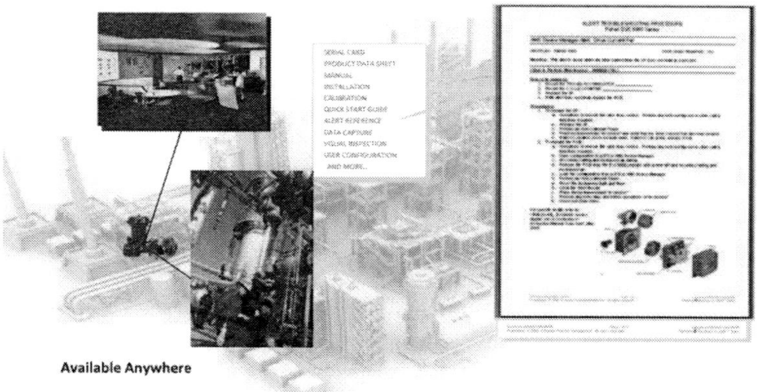

Photo Ref: Emerson Process Management

Move the data to the expert, not the expert to the data!

AIChE
Southwest Process Technology Conference

13

8th AIChE Southwest Process Technology Conference | October 6-7, 2016 | Galveston, TX

AFPM
American
Fuel & Petrochemical
Manufacturers

Making Sense And Cents From Data - Where Is The Payback?

Plant Operating Objectives - The Four Zero's

- ***Safety*** **– the goal is <u>zero serious safety incidents</u>**
- ***Sustainability*** **– the goal is <u>zero significant environmental incidents, excess energy use and excess waste</u>**
- ***Reliability*** **– the goal is <u>zero unscheduled downtime</u>**
- ***Financial*** **- the goal is <u>zero lost profit opportunities</u>**

How can Predictive Analytics support these objectives?

15

8th AIChE Southwest Process Technology Conference | October 6-7, 2016 | Galveston, TX

Plant Decision Cycle

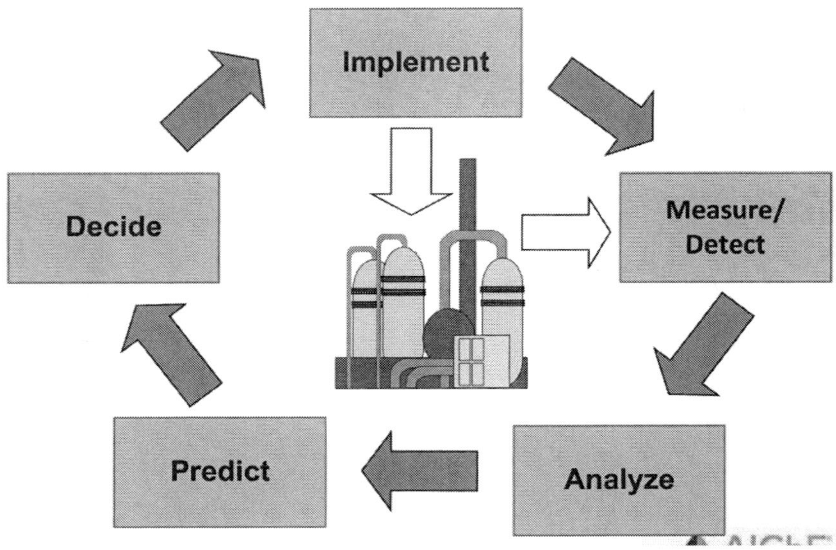

To have a financial impact predictive analytics has to improve this cycle – reduce delays, reduce uncertainty, etc.!

What Impact Can Predictive Analytics Have?

- **Safety**
  - Avoiding incidents through early detection of potential hazardous situations
- **Availability/ Reliability**
  - Anomaly detection – identifying precursor events to unscheduled equipment outage or problems
  - Performance monitoring – detecting loss of process/ equipment performance before it impacts production capacity
- **Sustainability**
  - Comparing current usage of resources such as energy to its expected usage under current conditions and determining possible causes of variation
- **Financial Optimization**
  - Detecting and dissecting complex interacting constraints on production
  - Determining reasons for product quality/ yield issues
  - Understanding patterns and relationships – developing statistical models that explain them

What are the Components of a Predictive Analytics Implementation?

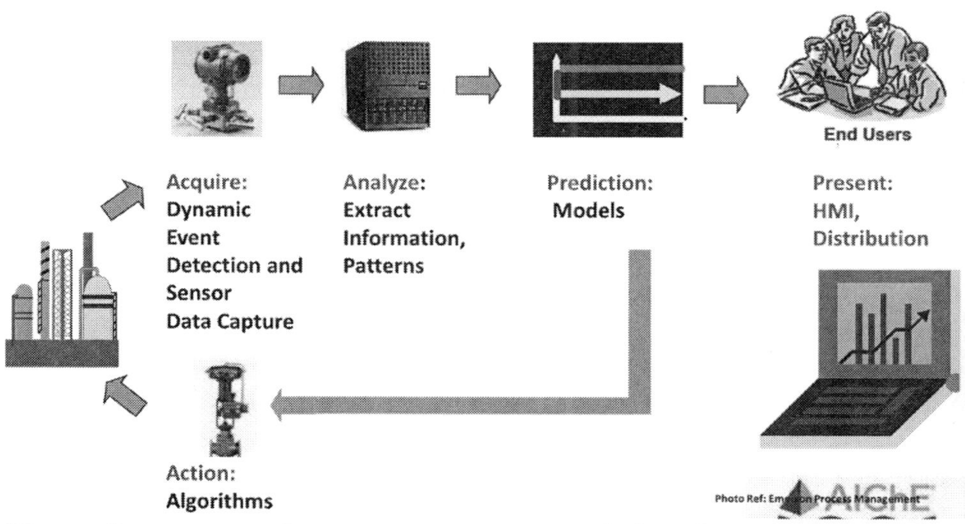

Move from reaction to events to anticipation and optimization

Major Predictive Analytics Techniques

- Numerical Data (with adjustment for time series dependence)
  - Develop Predictive Model
    - Linear Models
      - Continuous Variables – *Generalized Linear Regression*
      - Binary Outcome (Yes/No) – *Logistics Regression*
      - Discrete Outcome (1,2,3,4,..) *Classification and Reference Trees (CART + Random Forest)*
    - Non-Linear Models
      - Continuous Variables - *Neural Nets*
      - Discrete Outcome (1,2,3,4,..) - *(CART + Random Forest)*
  - Detect Anomaly
    - *Principal Component Analysis - PCA*
- Textual Data
  - *Textual Analytics*
- Spectral Data
  - *Nearest Neighbor*
  - *Regression*
- Video/ Images
  - *Neural Nets*

19

8th AIC

AFPM
American
Fuel & Petrochemical
Manufacturers

Typical Analytics Project

Typical Analytics Project Steps

1. Problem Definition – How will results be used to improve business performance
2. Data Location, Interpretation and Acq
3. Data Cleaning, Consolidation and Transformation
4. Model Building and Evaluation
5. Model Deployment, Monitoring and Updates

Often more than 50% of project hours

Plant Data Typical Questions

- Data
  - What data should we collect?
  - How should we store it? Particularly non-numeric data?
  - How should we search and retrieve data of interest?
  - How should we get the best data in the right hands?
  - How should we then act on that data?
  - How do we do this quickly and seamlessly with as little investment as possible?

- Platforms and Tools
  - What infrastructure should I use?
  - What tools should I use?
  - Should I store on-premise or in the cloud?
  - Should I use open source?
  - How do I scale?

Cleaning Data – Outlier Detection

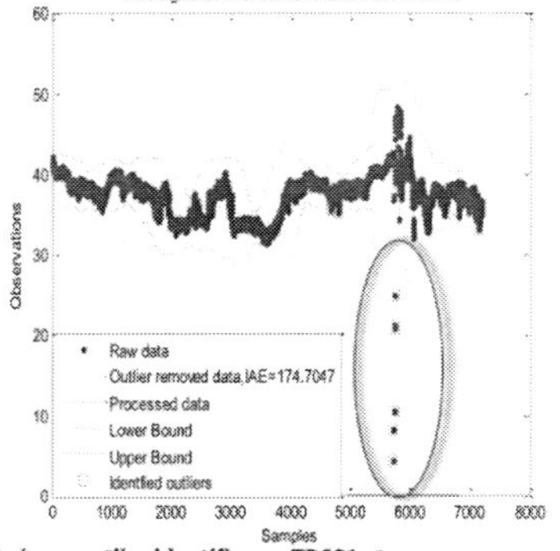

3-sigma outlier identifier on TD601 steam output data (moving window size=n/20=360)

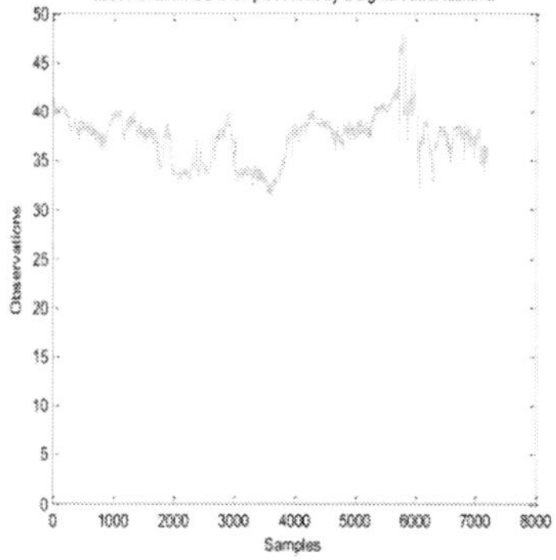

TD601 steam output data processed by 3-sigma outlier identifier(moving window size=n/20=360)

- Compressed data
- Missing data
- Outliers
- Noisy data

THE UNIVERSITY OF TEXAS AT AUSTIN
Cockrell School
McKetta Department of Chemical Engineering

Time Relationships Matter

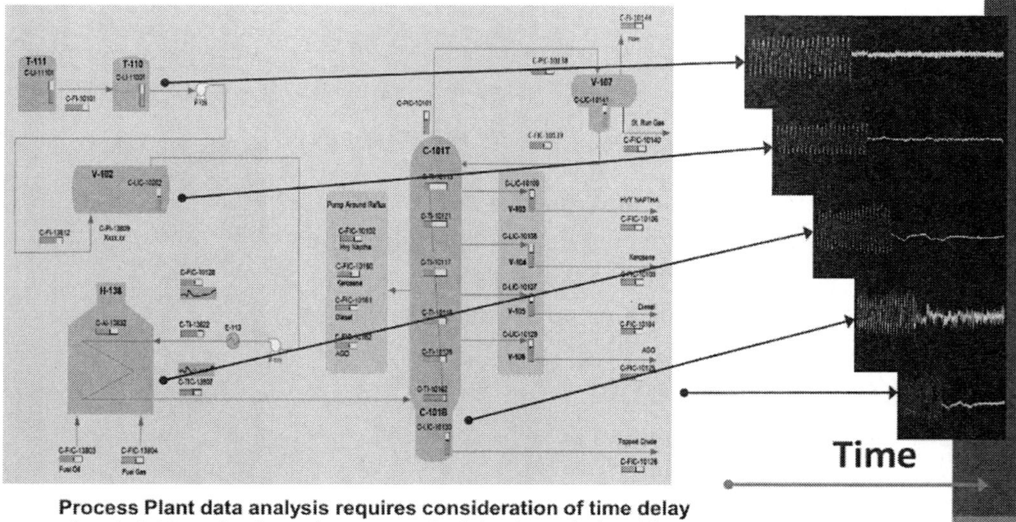

Process Plant data analysis requires consideration of time delay of material transfer through process. In detecting relationships, individual variable time delays have to be identified along with the correlation/ causation impacts – problem dimensionality becomes much higher.

AIChE
Southwest Process Technology Conference

Selecting Parameters And Time Delays

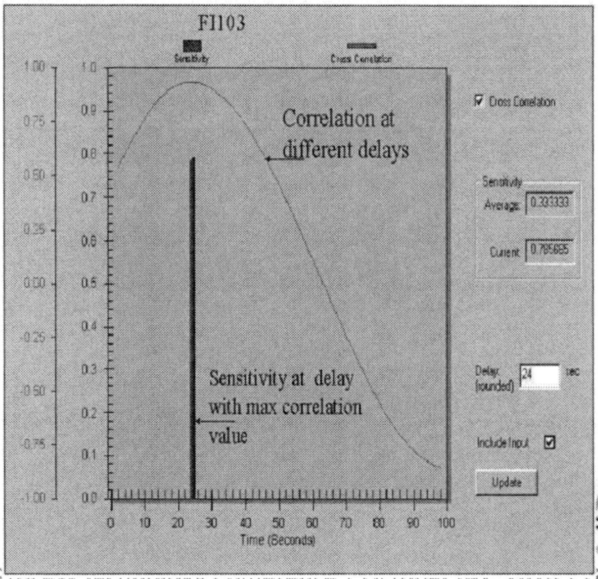

Develop and Evaluate Models

Utilize Block Diagramming Environment to develop and evaluate models

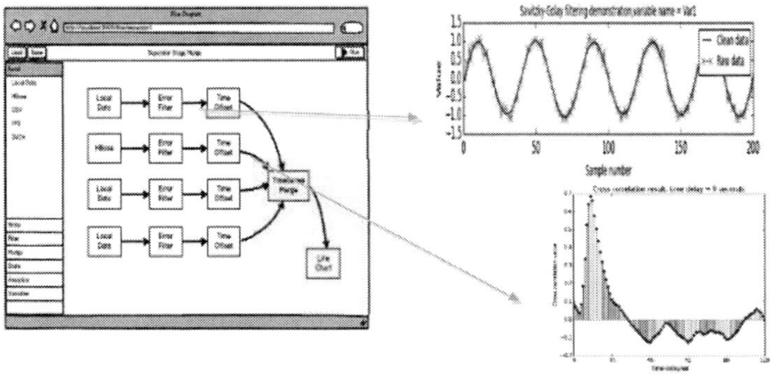

Interact with your model** **visualizing results at each stage of the data processing, when you are done,** **deploy the model** **for run-time execution

Visualizing Predictions

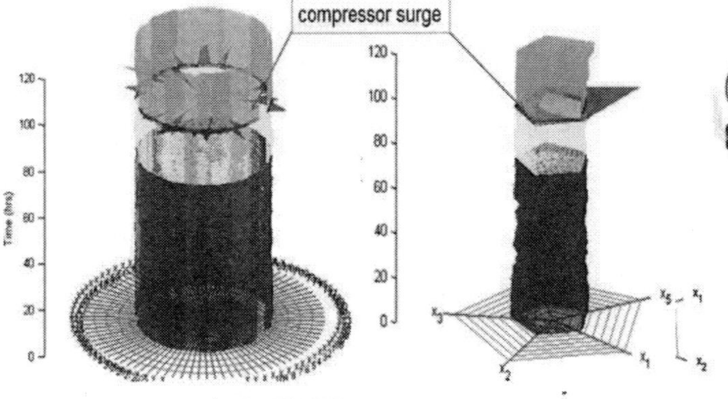

Inselberg, Comput. Stat., 1998

1 to 4 days before surge – Blue
1 day before surge – Yellow
2 hours before surge – Purple
Post surge - Green

Center for Operator Performance
An Industry/University Center for Excellence

THE UNIVERSITY OF TEXAS AT AUSTIN
McKetta Department of Chemical Engineering

AIChE
Southwest Process Technology Conference

Reference: Dunia, R; G. Rochelle; T.F. Edgar; M. Nixon; "Multivariate Modeling of a carbon dioxide removal process"; Computers & Chem Eng;60 (2014); pp. 381-395

AFPM
American Fuel & Petrochemical Manufacturers

Case Studies

Case Study – Predictive Analytics – Waste Water Plant

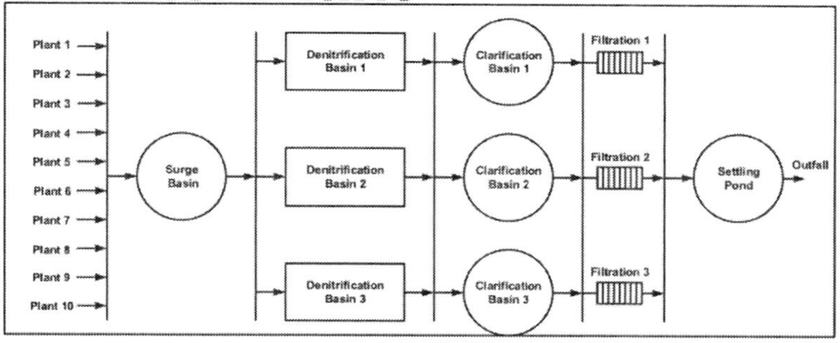

- Multiple effluent NH3 spikes in outfall violating consent decree and incurring fines
- Residence time of the treatment plant >2 weeks making issue ID difficult
- More than 200 measured variables to be considered as possible predictors

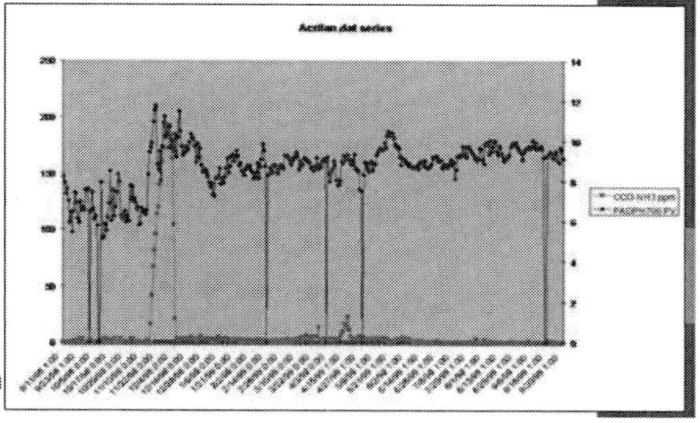

Case Study – Predictive Analytics – Waste Water Plant

- 5 years data analyzed for correlations with simultaneous identification of time lag
- Identification of strong correlation with pH of effluent from one of the plants – large drop in pH preceded excursion in denitrification basin by one day
- Vessel cleaning correspondence with pH drop

AIChE
Southwest Process Technology Conference

Case Study – Predictive Analytics - Maintenance

- **Significance**
  - Sipchem Jubail site is considered a world-class complex with 4 large integrated plants
  - Successful implementation of complete integrated maintenance program
  - Built on installed base to create multi-phase asset management program
- **Project Scope**
  - Implementation of Asset Management including Reliability Centered Maintenance on 6,000 assets
  - Uses predictive analytics and diagnostics from field instrumentation
- **Audited Value**
  - 12% decrease in maintenance costs
  - 2% increase in Plant availability

Saudi International Petrochemical Company (Sipchem) actively develops and invests in petrochemical and chemical industries.

Sources: Abdullah Al Ghamde; John E. Hill,; "*Saudi International Petrochemical Company (Sipchem) Reliability Process,*" Paper 01A241; 2010 Emerson Global User Exchange; Sudheer Prabhakaran; John E. Hill,;"*Generating Actual Savings with Sipchem's Reliability Program,*" Paper 1B-2487; 2012 Emerson Global User Exchange

Summary

- Predictive analytics is an evolving technology with many potential applications in the process industries
- Implementation of Predictive Analytics in the process industries has distinctive requirements/ issues due to the time series character and data
- Current applications in process fault detection, availability, safety and optimization have proven value – there will be more in the future

Thank You For Attending. Questions?

33

8th AIChE Southwest Process Technology Conference | October 6-7, 2016 | Galveston, TX

Automatically Detecting Operating Procedures from Raw Process Data using Filtering, Machine Learning, and Clustering Methods

Richmond, P. C., Lamar University

Originally presented on: 10/6/2016 10:20:00 - 10:45:00

A data analysis framework is proposed to identify unlogged recurring process transitions from raw process data. Transitions are a ubiquitous part of any continuous process occurring every time an operating procedure includes opening valves, starting pumps, or taking equipment in or out of service. However, since many of these procedural steps are accomplished manually there are no data logs that can be used with traditional data mining techniques. In this work we investigate using data filters and machine learning techniques to identify recurring transitions from process data. The information extracted can then be used to identify clusters of data and detect similar patterns representing recurring process transitions thus automatically creating operating procedure data logs.

Managing the Lifecycle of Independent Protection Layers

Patrick Fisher, Provenance Consulting, October 6, 2016

AIChE Southwest Process Technology Conference

8th AIChE Southwest Process Technology Conference | October 6-7, 2016 | Galveston, TX

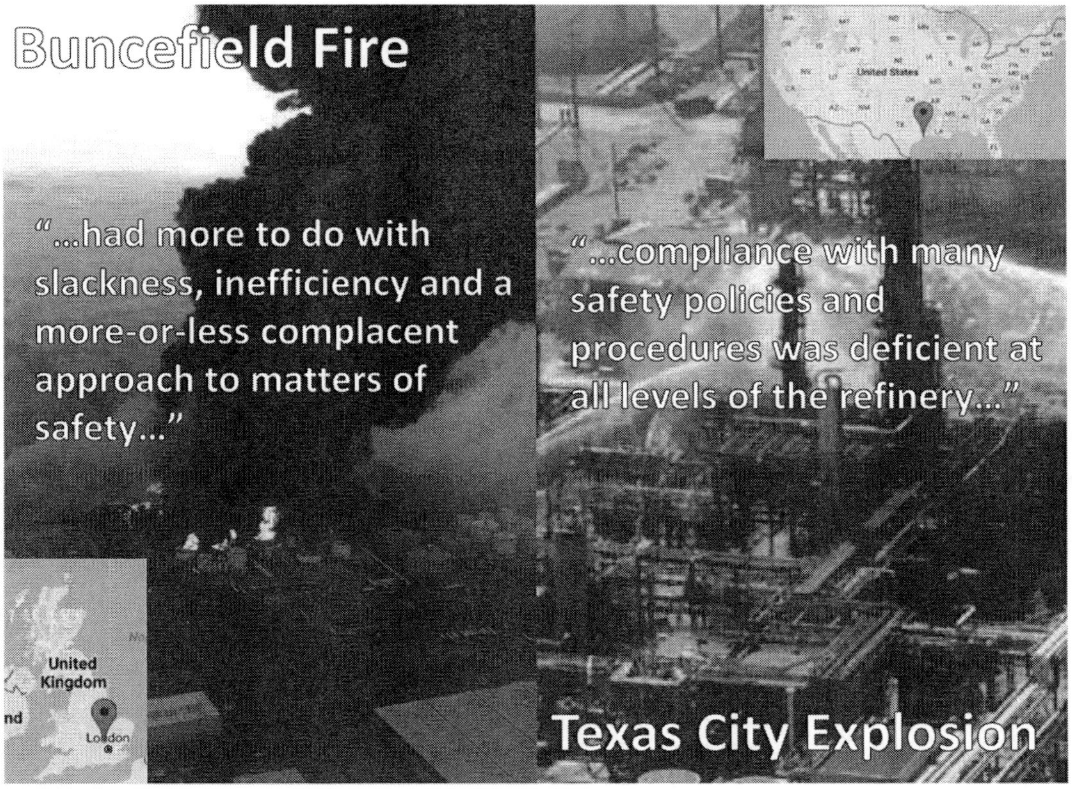

Overview > What is an IPL? > Managing IPLs > In practice > Wrap up

Overview

What is an IPL?

When is a safeguard an IPL?

What makes a good IPL?

What criteria of IPLs need to be managed?

CCPS seven core attributes of IPLs

What needs to be documented to meet these criteria?

What does the management of IPLs look like in practice?

How does the documentation get managed?

How can software help manage the documentation?

AIChE Southwest Process Technology Conference

8th AIChE Southwest Process Technology Conference | October 6-7, 2016 | Galveston, TX

Overview > What is an IPL? > Managing IPLs > In practice > Wrap up

What is an IPL?

"A device, system, or action that is, capable of preventing a scenario from proceeding to its undesired consequence, **independent of the initiating event or the action of any other layer of protection associated with the scenario.** <u>*The effectiveness and independence of an IPL must be auditable*</u>"

Layer of Protection Analysis: Simplified Process Risk Assessment, CCPS 2001

"A device, system, or action that is capable of preventing a scenario from proceeding to the undesired consequence **without being adversely affected by the initiating event or the action of any other protection layer associated with the scenario**"

Guidelines for Initiating Events and Independent Protection Layers in Layer of Protection Analysis, CCPS 2015

AIChE Southwest Process Technology Conference

8th AIChE Southwest Process Technology Conference | October 6-7, 2016 | Galveston, TX

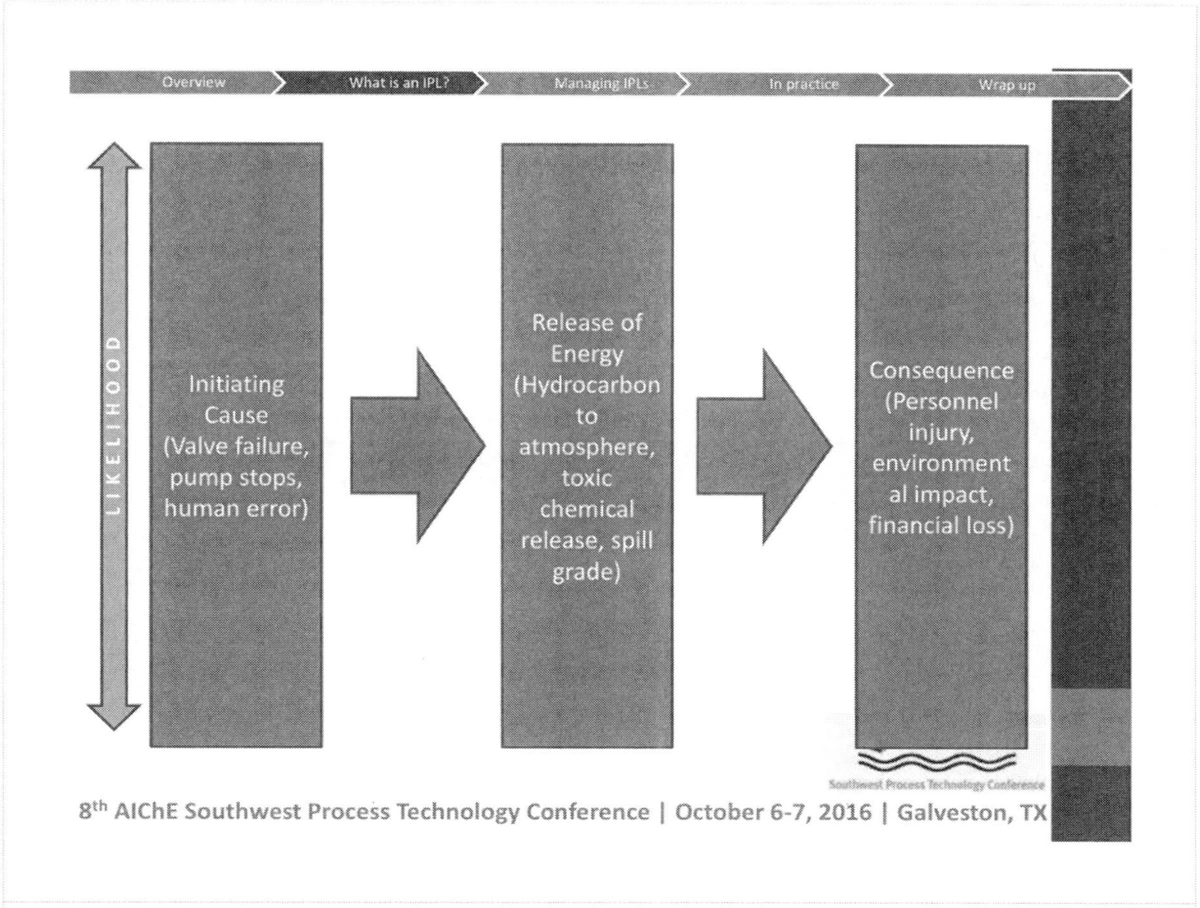

Overview
What is an IPL?
Managing IPLs
In practice
Wrap up
LIKELIHOOD
Initiating Cause (Valve failure, pump stops, human error)
Release of Energy (Hydrocarbon to atmosphere, toxic chemical release, spill grade)
Consequence (Personnel injury, environment al impact, financial loss)
Southwest Process Technology Conference
8th AIChE Southwest Process Technology Conference | October 6-7, 2016 | Galveston, TX

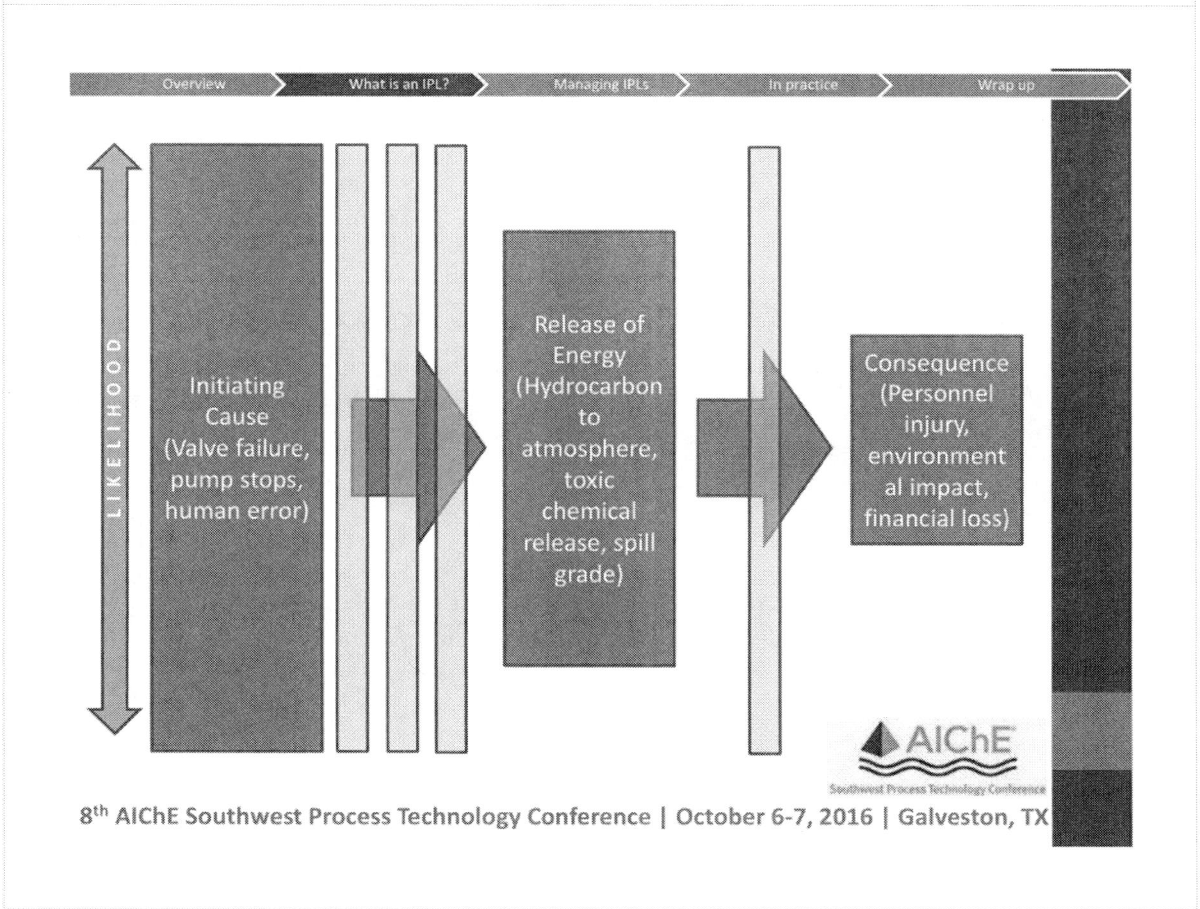

Overview
What is an IPL?
Managing IPLs
In practice
Wrap up
LIKELIHOOD
Initiating Cause (Valve failure, pump stops, human error)
Release of Energy (Hydrocarbon to atmosphere, toxic chemical release, spill grade)
Consequence (Personnel injury, environment al impact, financial loss)
AIChE
Southwest Process Technology Conference
8th AIChE Southwest Process Technology Conference | October 6-7, 2016 | Galveston, TX

Safeguards vs IPLs

Any safeguard identified in the HAZOP can potentially qualify as an IPL.

Independent | Effective | Auditable

Independent Protection Layers

Independence

Plugging

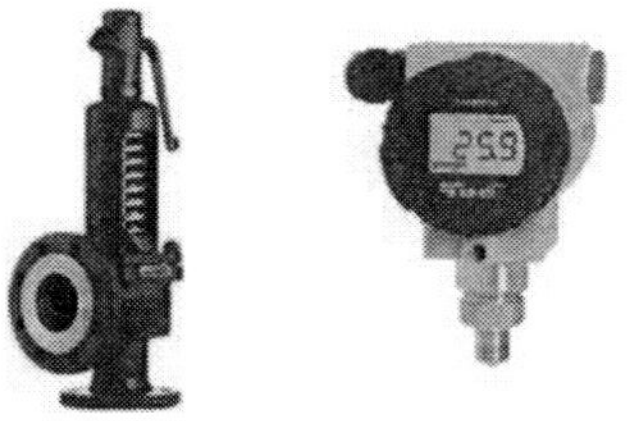

Mechanical integrity

"IPLs should be sufficiently independent such that the degree of interdependence is not statistically significant"

Guidelines for Initiating Events and Independent Protection Layers in Layer of Protection Analysis, CCPS 2015

Overview > What is an IPL? > Managing IPLs > In practice > Wrap up

Independence

A safeguard from the HAZOP can be considered independent if

- It has non-significant interdependence with initiating event
- It has non-significant interdependence with other IPLs
- It is not adversely affected by the initiating event
- It is not adversely affected by other IPLs

Independence

AIChE Southwest Process Technology Conference

8th AIChE Southwest Process Technology Conference | October 6-7, 2016 | Galveston, TX

Overview > What is an IPL? > Managing IPLs > In practice > Wrap up

Effective

*"A device, system, or action that is **capable of preventing a scenario from proceeding to the undesired consequence** without being adversely affected by the initiating event or the action of any other protection layer associated with the scenario."*

Guidelines for Initiating Events and Independent Protection Layers in Layer of Protection Analysis, CCPS 2015

AIChE Southwest Process Technology Conference

8th AIChE Southwest Process Technology Conference | October 6-7, 2016 | Galveston, TX

Effective

Effective Relief Valve

- Sized for scenario
- Set at the right pressure
- In the right service (plugging, clean)
- Regularly tested or replaced

Effective Alarm

- Would detect the scenario
- Set with sufficient time to act
- Reliable instrument
- Operator action is clear

Effective

A HAZOP safeguard can be considered effective if

- It is functional enough to prevent the scenario without any outside interference
- It has sufficient integrity to provide the required risk reduction
- It is reliable enough to provide sufficient risk reduction between tests

Effectiveness

Auditable

"The effectiveness and independence of an IPL must be auditable."

Layer of Protection Analysis: Simplified Process Risk Assessment, CCPS 2001

Auditable Relief Valve

Can be tested on a regular basis

Can be checked for service (clean, plugging)

Auditable Alarm

Can be tested on a regular basis
The intended operator response is documented
Actual operator responses are documented

Safeguards to IPLs

	Independent	Effective	Auditable
Procedures	Moderate	Easy	Moderate
Alarms	Easy	Easy	Easy
Trips	Easy	Easy	Easy
Relief Valves	Easy	Easy	Easy
Operator rounds	Easy	Hard	Moderate
Signage	Hard	Hard	Easy
Samples	Easy	Hard	Moderate
Emergency response plans	Moderate	Moderate	Moderate
Post-release (deluge, etc)	Easy	Easy	Moderate

Safeguards to IPLs

	Independent	Effective	Auditable
Procedures	Moderate	Easy	Moderate
Alarms	Easy	Easy	Easy
Trips	Easy	Easy	Easy
Relief Valves	Easy	Easy	Easy
Operator rounds	Easy	Hard	Moderate
Signage	Hard	Hard	Easy
Samples	Easy	Hard	Moderate
Emergency response plans	Moderate	Moderate	Moderate
Post-release (deluge, etc)	Easy	Easy	Moderate

AIChE Southwest Process Technology Conference

What is an IPL?

An IPL is a safeguard which we choose to focus our attention on because it is easier to quantify, qualify, and verify the amount of safety protection provided

	Independent	Effective	Auditable
Procedures	Moderate	Easy	Moderate
Alarms	Easy	Easy	Easy
Trips	Easy	Easy	Easy
Relief Valves	Easy	Easy	Easy
Emergency response plans	Moderate	Moderate	Moderate
Post-release (deluge, etc)	Easy	Easy	Moderate

Southwest Process Technology Conference

What comes after identifying IPLs?

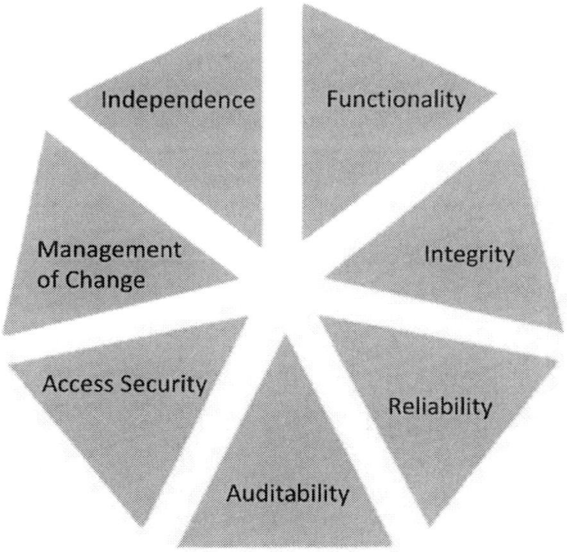

There are seven core attributes to an IPL which need to be managed through the IPL lifecycle

Guidelines for Safe and Reliable Instrumented Protective Systems (CCPS 2007), Guidelines for Initiating Events and Independent Protection Layers in Layer of Protection Analysis, CCPS 2015

Independence

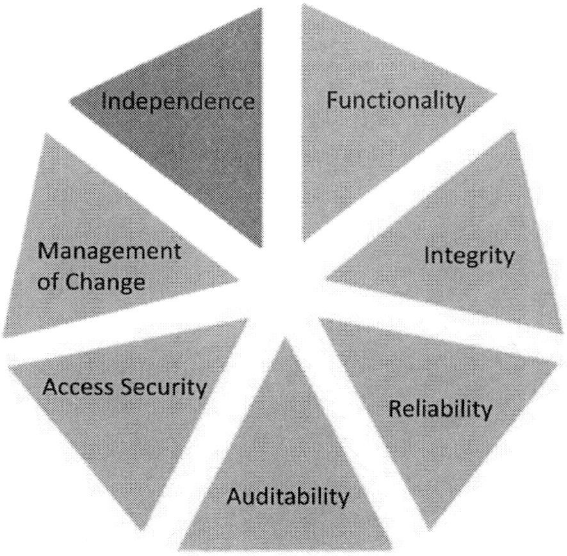

For each IPL, document all associated equipment, personnel, and control systems.

For each initiating event, document the potential failures, including individual pieces of equipment and personnel

Independence

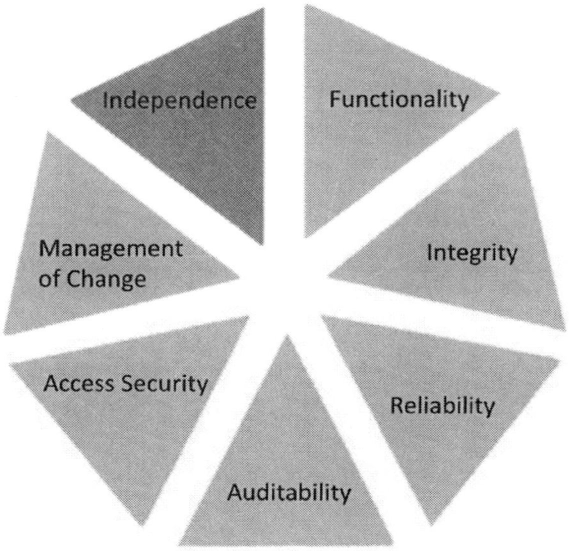

If any of the elements are common, then ensure that the likelihood/probability of failure of the common element is less than the total PFD of the combined IPLs/IEFs

Functionality

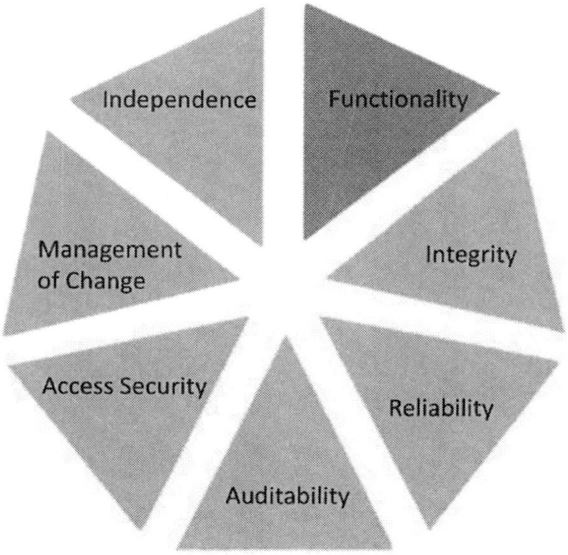

Document all relevant requirements for the IPL, including:

Instrument

Set Point

Logic Solver System

Final element (e.g. valve or operator)

Time required to respond

Expected result of IPL activation

Overview | What is an IPL? | Managing IPLs | In practice | Wrap up

Integrity

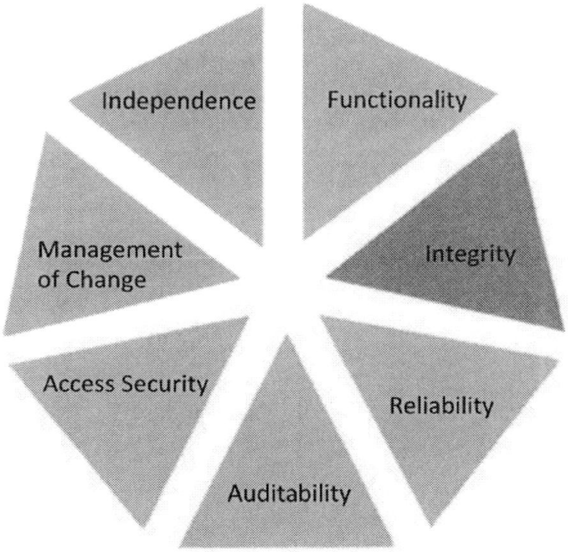

For each cause-consequence pair, document the PFD of the IPL. This is the maximum or worst) PFD that can be accepted from the IPL.

Integrity is limited to the weakest element

AIChE Southwest Process Technology Conference

8th AIChE Southwest Process Technology Conference | October 6-7, 2016 | Galveston, TX

Overview | What is an IPL? | Managing IPLs | In practice | Wrap up

Reliability

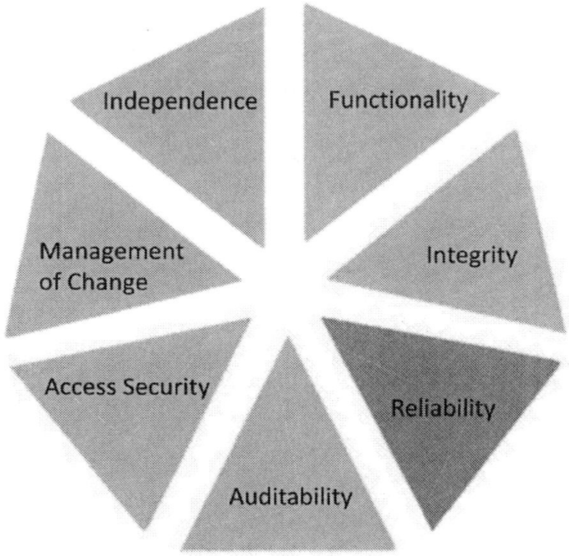

Document the design conditions expected, and include a plan for mitigation if the IPL is not available

The mitigation plan ensures reliability of the system even when the primary IPL is not available

AIChE Southwest Process Technology Conference

8th AIChE Southwest Process Technology Conference | October 6-7, 2016 | Galveston, TX

Auditability

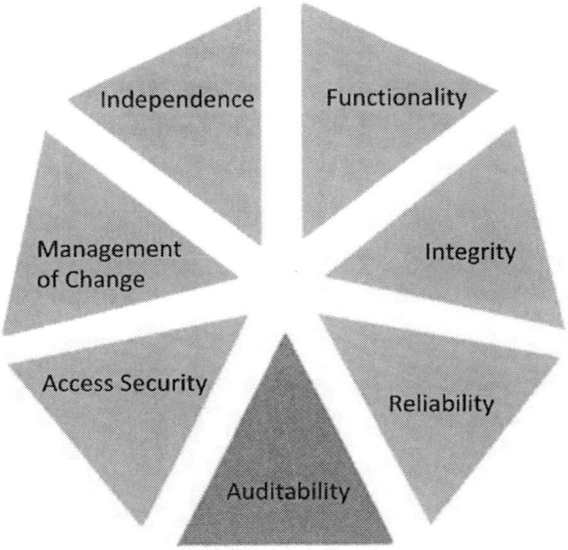

Prepare an audit plan which includes how to verify the independence and effectiveness of the IPL.

At a minimum, the independence and effectiveness should be reviewed along with the PHA cycle (5 years)

AIChE Southwest Process Technology Conference

Access Security

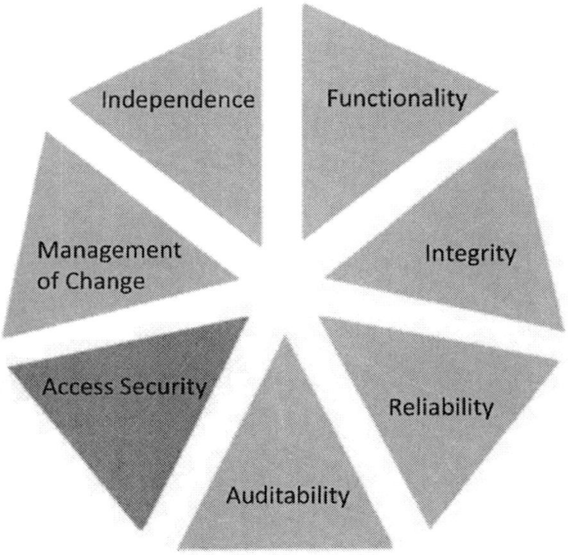

IPLs must have physical and/or administrative controls to reduce unauthorized system changes

This includes malicious and accidental change

Document the controls in place so that they can be audited

AIChE Southwest Process Technology Conference

Management of change

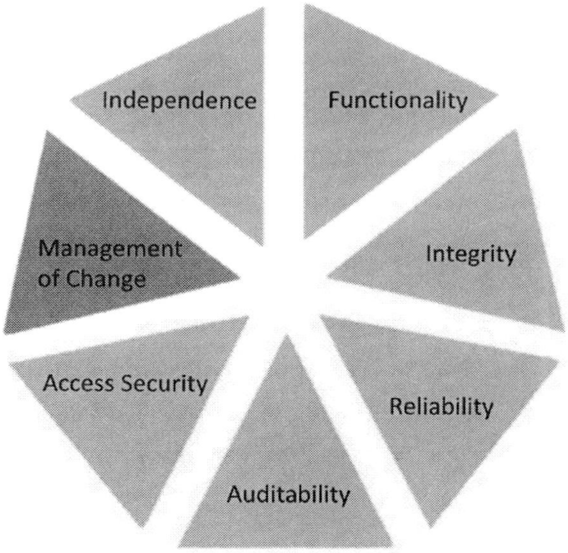

Flag any change that may affect independence or effectiveness

If any element of an IPL changes, all other associated IPLs and Initiating events should also be reviewed for impacts.

Overview > What is an IPL? > Managing IPLs > In practice > Wrap up

What comes after identifying IPLs?

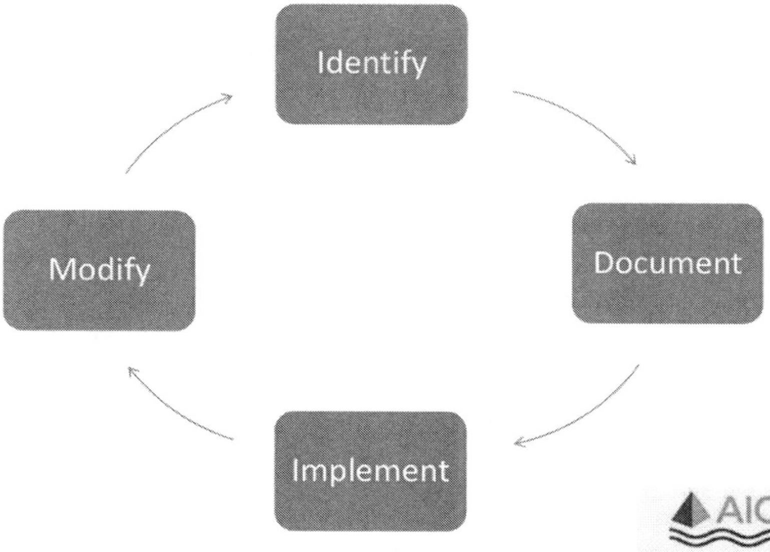

8th AIChE Southwest Process Technology Conference | October 6-7, 2016 | Galveston, TX

Identifying and Documenting IPLs

IPL Description

PI-101 high pressure alarm, set at 100psig, will alarm through the control system and the Area B board operator will respond within 8 minutes by closing the fuel gas valve HV-102 before the pressure reaches the V-104 MAWP of 120psig.

Source

2016 Area B HAZOP/LOPA Revalidation, Nodes 2 and 3.

Components

Instrumentation	Control System	Final Elements	Personnel
PT 101	I/O Card 856 CPU 927 I/O Card 639 Logic block 827 Screen 21	HV-102 HV-102 Solenoid	Area B Board Operator Area A/B Instrumentation tech

AIChE Southwest Process Technology Conference

Overview > What is an IPL? > Managing IPLs > In practice > Wrap up

Identifying and Documenting IPLs

Identified Hazards

PHA Ref.	Consequence	Cat.	Severity	Risk	PFD
2.6.1.3	Potential overpressure of V-104 by greater than 2x MAWP (120psig). Potential for multiple gasket failures and loss of containment of hydrocarbon with potential fire and personnel injury.	Safety	Permanent Disability	2 – Moderate	0.1
3.4.3.1	Potential overpressure of V-104 by greater than 2x MAWP (120psig). Potential for multiple gasket failures and loss of containment of hydrocarbon with potential fire and personnel injury.	Safety	Permanent Disability	1 - Low	0.1

Process Safety Time

PHA Ref.	Time from detection to release	Expected time to action	Expected Result
2.6.1.3	15 minutes	8 minutes	Tower temperature TI-101 will drop and PI-101 will no longer read high.
3.4.3.1	18 minutes	8 minutes	Tower temperature TI-101 will drop and PI-101 will no longer read high.

AIChE Southwest Process Technology Conference

8th AIChE Southwest Process Technology Conference | October 6-7, 2016 | Galveston, TX

Identifying and Documenting IPLs

Identified Hazards

PHA Ref.	Consequence	Cat.	Severity	Risk	PFD
2.6.1.3	Potential overpressure of V-104 by greater than 2x MAWP (120psig). Potential for multiple gasket failures and loss of containment of hydrocarbon with potential fire and personnel injury.	Safety	Permanent Disability	2 – Moderate	0.1
3.4.3.1	Potential overpressure of V-104 by greater than 2x MAWP (120psig). Potential for multiple gasket failures and loss of containment of hydrocarbon with potential fire and personnel injury.	Safety	Permanent Disability	1 - Low	0.1

Process Safety Time

PHA Ref.	Time from detection to release	Expected time to action	Expected Result
2.6.1.3	15 minutes	8 minutes	Tower temperature TI-101 will drop and PI-101 will no longer read high.
3.4.3.1	18 minutes	8 minutes	Tower temperature TI-101 will drop and PI-101 will no longer read high.

AIChE Southwest Process Technology Conference

Identifying and Documenting IPLs

Initiating Event: FV-306 malfunctions closed or manual block valves closed

Instrumentation	Control System	Mechanical Elements	Personnel
FT-306 PT-100	I/O Card 855 CPU 927 I/O Card 656 Logic block 821 Screen 21	FV-306 FV-306 Solenoid	Area B Board Operator Area A/B Instrumentation tech Area B field operator

Common elements notes: CPU is common between IPL and IE, however the frequency of CPU failure is acceptable for this condition (0.01). Area B board operator is common between IPL, however the frequency of human failure (low stress, non-fatigued) is acceptable for this condition (0.01).

Initiating Event: Manual block valve on unit edge misaligned closed.

Instrumentation	Control System	Mechanical Elements	Personnel
N/A	N/A	Manual Valve on Line 837 for vapor overhead between Area A and B	Area B field operator Area A field operator

Common elements notes: No common elements between IPL and IE.

Related IPLs: Relief valve RV-902 set at 120psig, sized for blocked outlet of V-104 relieves to Area B Flare

Instrumentation	Control System	Final Elements	Personnel
N/A	N/A	RV-902 CSO Valve 5426 CSO Valve 3028 Area B Flare	Area B Maintenance tech

Common Element note: Potential for process plugging for both PI and Relief Valve. Both are in clean service, and probability is considered very low (<0.001)

AIChE Southwest Process Technology Conference

Managing a company full of IPLs

ARTS Asset & Records Traceability System

IPL Health by Facility

Facility	IPL Health	PSI Health
Houston	84% - Poor	97% - Excellent
LA	97% - Excellent	93% - Good
Minneapolis	92% - Moderate	88% - Poor
Denver	94% - Good	98% - Excellent

Managing a company full of IPLs

ARTS Asset & Records Traceability System

IPL Health by Unit - Houston

Unit	Independence	Effectiveness	Auditing	Access Security
Area A	100% - Excellent	100% - Excellent	95% - Good	100% - Excellent
Area B	65% - Very Poor	75% - Poor	80% - Poor	100% - Excellent
Area C	95% - Good	98% - Excellent	97% - Good	100% - Excellent
Area D	100% - Excellent	100% - Excellent	99% - Good	100% - Excellent

Overview > What is an IPL? > Managing IPLs > In practice > Wrap up

Managing a company full of IPLs

ARTS Asset & Records Traceability System

IPL Health by IPL – Houston Area B

Protection Layer	Independence	Effectiveness	Auditing	MOC Detection	Access Security
IPL 1 – LAH-203...	Verified	Verified	Up to Date	Flagged	In place
IPL 2 – PAH-101...	Unverified	Not meeting specifications	Testing Overdue	Flagged	In place
IPL 3 – FAL-104...	Partially Verified	Unverified	Up to Date	Not Flagged	In place
IPL 4 – PAL-308...	Verified	Partially Verified	Testing missing	Flagged	In place

Overview > What is an IPL? > Managing IPLs > In practice > Wrap up

Managing a company full of IPLs

ARTS Asset & Records Traceability System

IPL Health by IPL – Houston Area B - IPL 2

IPL Description

PI-101 high pressure alarm, set at 100psig, will alarm through the control system and the Area B board operator will respond within 8 minutes by closing the fuel gas valve HV-102 before the pressure reaches the V-104 MAWP of 120psig.

Testing

Component	Design Conditions	PFD	Test Interval	Actual Conditions	Last Test
PT 101	Clean, low vibration	0.1	3 years	Clean, low vibration	3.5 years
I/O Card 856	Instrument enclosure	0.1	3 years	Instrument enclosure	3.5 years
CPU 927	Instrument enclosure	0.01	3 years	Instrument enclosure	2 years
I/O Card 639	Instrument enclosure	0.1	3 years	Instrument enclosure	3.5 years
Logic block 827	N/A	0.1	3 years	N/A	3.5 years
Screen 21	Meets HF standard RS	0.01	1 year	Meets HF standard RS	6 months
HV-102	Clean, low vibration	0.1	3 years	Clean, high vibration	2 years
HV-102 Solenoid	Clean, low vibration	0.1	3 years	Clean, high vibration	2 years
Area B Board operator	Low stress, non-fatigued	0.01	1 year refresher	Low stress, non-fatigued	6 months
Area A/B Instrument Tech	Low stress, non-fatigued	0.1	1 year refresher	Low stress, non-fatigued	6 months

8th AIChE Southwest Process Technology Conference | October 6-7, 2016 | Galveston, TX

Recording IPL performance

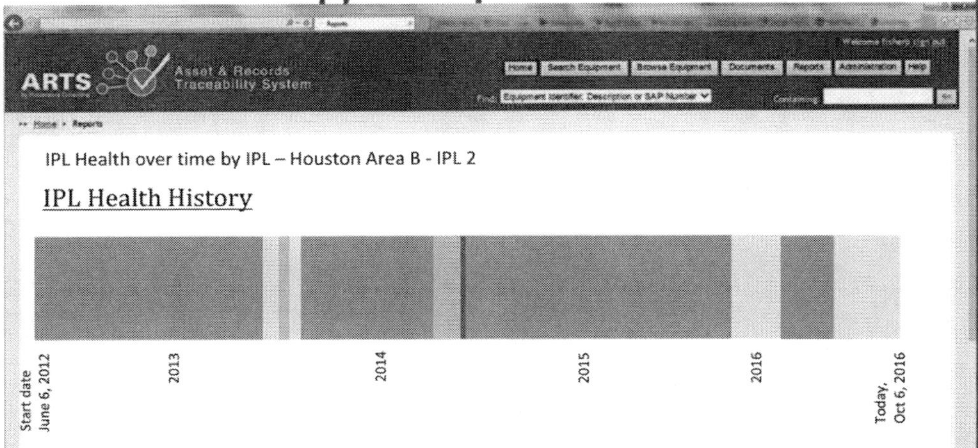

	Green (Excellent)		Yellow (Good)		Orange (Moderate)		Red (Poor)	
3 years	1029 days,	94%	43 days,	4%	22 days,	2%	1day ,	<1%
1 year	350 days,	96%	15 days,	4%	0 days,	0%	0 days,	0%

Wrapping up

What is an IPL?

An IPL is a safeguard that can be proven to be effective

How do we prove an IPL is reliable?

Document and implement the seven criteria for IPLs

How do you manage an IPL lifecycle?

Monitor the documentation and implementation of the IPLs.

AIChE Southwest Process Technology Conference

United Kingdom
London
United States

The Evolution of Process Safety: Current Status and Future Direction

Mannan, S., Texas A&M University

Originally presented on: 10/6/2016 13:30:00 - 14:00:00

The advent of the industrial revolution in the 19th century increased the volume of manufactured goods and enriched the quality of life for society as a whole. However, industrialization was also accompanied by new manufacturing and complex processes that brought about the use of hazardous chemicals and difficult-to-control operating conditions. Moreover, human-process equipment interaction plus on-the-job learning resulted in further undesirable outcomes and associated consequences. These problems gave rise to many catastrophic process safety incidents that resulted in thousands of fatalities and injuries, losses of property, and environmental damages. These events led eventually to the necessity for a gradual development of a new multidisciplinary field, referred to as process safety. From its inception in the early 1970s to the current state of the art, process safety has come to represent a wide array of issues, including safety culture, process safety management systems, process safety engineering, loss prevention, risk assessment, risk management, and inherently safer technology. Governments and academic/research organizations have kept pace with regulatory programs and research initiatives, respectively. Understanding how major incidents impact regulations and contribute to industrial and academic technology development provides a firm foundation to address new challenges, and to continue applying science and engineering to develop and implement programs to keep hazardous materials within containment. Here the most significant incidents in terms of their impact on regulations and the overall development of the field of process safety are described.

Conduct of Operations - "Walk the Line!"

Abrahamson, D., Abrahamson Consulting LLC

Originally presented on: 10/6/2016 14:00:00 - 14:30:00

Conduct of Operations is one of the 20 Risk Based PSM elements from the Center for Chemical Process Safety (CCPS) - it is about running a "Tight Ship" or "Walk the Line". Conduct of Operations is about performing tasks according to expectations, policies and procedures and behaving in a safe and professional manner to reduce variability, reduce human error, increase safety and enhance effectiveness. Specific examples will be discussed on what to include in a Conduct of Operations management system to institutionalize the pursuit of excellence in the performance of everyday tasks and to minimize variations in performance. Examples on how to incorporate Conduct of Operations activities into your operations such as: Process Safety Officer, Operations Shift Change, Maintenance-Operations Hand-Over, Verification after Maintenance, Communications, Inhibiting Safety Systems, Labeling, Housekeeping, Lighting, Tool Management, Control of Access to Unit and Safe Minimum Staffing.

Can Your Operators Respond Quickly Enough to Safety Alarms?

White, J., SIS-Tech

Originally presented on: 10/6/2016 14:30:00 - 15:00:00

Plant reliability and safety is improved when there is confidence that correct operator response will occur. Unfortunately, a disproportionate share of process safety incidents has occurred where the required actions did not happen as planned or happened too late. The effectiveness of a safety alarm is limited primarily by the probability that the operator follows a defined course of action in a timely manner in response to the alarm presentation. The operator requirements must be defined both in terms of needed action(s) and response time. This paper will discuss some of the challenges in defining the operator requirements and in validating the required actions and timeliness. It will then propose some approaches to meeting those challenges.

Comparison of Risk Management Practices Amongst Companies

Devlin, C., Celanese Corporation
McCollum, B.,

Originally presented on: 10/6/2016 15:00:00 - 16:00:00

This session is a panel with members from Celanese, Dow and one additional panelist to be determined. Each of the panelists will take time to describe their Risk Management programs, which includes their approach to PHA and higher risk assessment techniques including LOPA, Quantitative Risk Analysis and Fault Tree Analysis. The prepared remarks by each of the panelists will be followed by a moderated Q&A session.

Diagnosing and Resolving Process Flow Maldistribution in a Packed Column
Pless, L., Tracerco
Originally presented on: 10/6/2016 13:30:00 - 13:55:00

Diagnosing and Resolving Process Flow Maldistribution

in a Packed Column

Lowell Pless[1], Nicholas Graham[2], Mason Bieker[2]Â and Rodney Thompson[1], (1) Tracerco, Pasadena, TX, (2) ChevronPhillips Chemical Co, Baytown, TX

An olefins column was having separation problems. A project team was assembled to first diagnose the source of the problem and then to recommend and implement a solution. First step taken by the project team was to use gamma scans to study the hydraulic performance of the column.

The column was configured with a top section of structured packing and a bottom section of one-pass trays. Initially it was not known what or where the potential problem was so both sections of the column were scanned at "normal" process conditions.

The gamma scan results, including tomography scans, indicated that the packed section was operating with some degree of process flow maldistribution while the trays appeared to be operating satisfactorily. Now the debate moved to whether the flow maldistribution through the packed bed was liquid (liquid distributor) related or vapor related.

The project team devised a novel approach to plant test whether the packing maldistribution was liquid related or vapor related. Gamma scans were used to ascertain the effects on the maldistribution from the plant test. The gamma scan results confirmed to the project team what then needed to be revamped in order to mitigate the columnâ??s separation problem. Some internal hardware was re-designed and later installed after which the column began to perform its intended separation.

This presentation will describe the plant test procedure and the thought processes based on the scan results, other process data and hardware design criteria. The process improvement plan that was designed and implemented will be discussed.

Improved Performance of NGL Fractionation with the Optimax (tm) Process

Bravo, J., Shell Global Solutions (US) Inc.
Suffridge, G.,

Originally presented on: 10/6/2016 13:55:00 - 14:20:00

Gas plant operations to recover Natural Gas Liquids (NGLs) are very sensitive to the composition of the residue or sales gas. When Ethane process are high the plants need to operate in an ethane recovery mode with little ethane lost to the residue gas which would be mostly methane. But in conditions where the ethane value is low, it is advantageous to reject the ethane with the methane and supply NGLs to fractionation that are heavier. This paper will describe how to configure a de-methanizer to achieve either of the two purposes without needing expensive revamps and preserving the flexibility to operate in either mode. The Proprietary Configuration is called the S & B Optimax process.

Dry Tray Pressure Drop of Sieve Trays Re-visted and Expanded

Summers, D. R., Sulzer Chemtech USA, Inc.
Cai, T., Fractionation Research, Inc.

Originally presented on: 10/6/2016 14:20:00 - 14:45:00

The Dry Tray Pressure Drop of Sieve Trays has been explored extensively over the decades. Both authors have presented papers on the topic in the past 10 years. The work of both authors complements each otherâ??s work and it is our belief that this document will finally "put to rest" this topic for eternity. We will present a new Dry Tray pressure drop correlation that is superior to all who have come before us and should become the foundation of other tray hydraulic parameters for Sieve Trays.

cMIST™: Novel, compact gas dehydration system

Ramkumar, S., ExxonMobil
Grave, E., ExxonMobil Upstream Research Company
Yeh, N. K., ExxonMobil Upstream Research Company
Originally presented on: 10/6/2016 14:45:00 - 15:10:00

cMISTâ?¢: Novel, compact gas dehydration system

Shwetha Ramkumar, Edward J. Grave, Norman K. Yeh

ExxonMobil Upstream Research Company

Objectives/Scope

The natural gas industry is increasingly pursuing compact and lower weight processing technology to meet the technological and economic demands of offshore, onshore, remote, and challenged gas processing. Increased modularity of process equipment and process intensification require innovative solutions. This paper describes ExxonMobil Upstream Research Company's (EMURC) new inline dehydration and separation technology meant to replace conventional TEG tower and associated separator vessels to meet pipeline dewpoint specifications. Compact Mass transfer and Inline Separation Technology (cMISTâ?¢) achieves this goal with significant reductions in weight and footprint.

Methods, Procedures, Process

EMURC is qualifying cMISTâ?¢ for upstream natural gas dehydration. The results of a field demonstration are discussed to establish the effectiveness and advantages of this novel technology. The technology was demonstrated using a skid-mounted test unit in the Woodford trend area of Oklahoma, USA at an XTO Energy production facility. cMISTâ?¢ was tested under a range of field conditions to examine the dehydration capacity, operability, and reliability of the technology.

Results, Observations, Conclusions

cMIST relies on state-of-the-art technological advances in contacting and inline separation to achieve pipeline specification dry gas and/or prevent hydrate formation with compact equipment. The technology is highly modular to allow for simple transportation into remote, challenged, or offshore environments and provide installation configuration flexibility. One case study showed a 50% weight reduction and 80% footprint reduction for the absorption system on an offshore platform dehydrating 700 MMscfd of gas. During EMURC's field demonstration and qualification program, cMIST achieved pipeline specification over a range of process conditions and during sustained operation. cMIST was demonstrated to have a broad application in gas dehydration.

DISTILLATION COLUMN DESIGN AND ANALYSIS

8TH AICHE SOUTHWEST PROCESS TECHNOLOGY CONFERENCE

OCTOBER 6-7, 2016 I GALVESTON, TX

CREATE AMAZING.

AGENDA

- Introduction
- Types of Columns
- Distillation Principles
- Distillation Design: Eight Practical Steps
- Useful Resources

DISTILLATION IS...

- a process in which a liquid or vapor mixture of two or more substances is separated into its component fractions of desired purity
- based on the fact that the vapor of a boiling mixture will be richer in the components that have lower boiling points
- a consumer of enormous amounts of energy in terms of cooling and heating
- a large contributor to plant operating costs, maybe more than 50%

BURNS & McDONNELL

3

AGENDA

- Introduction
- Types of Columns
- Distillation Principles
- Distillation Design: Eight Practical Steps
- Factors Affecting Operation
- Useful Resources

BURNS & McDONNELL

4

TYPES OF DISTILLATION COLUMNS

- Packed Bed Columns
  - Used often for absorption and distillation of vapor-liquid mixtures
  - Liquid flows downward through the packing
  - Vapor flows upward through the packing
- Advantages
  - Cost efficient
  - Lower pressure drop
  - Good for thermally sensitive liquids
- Disadvantages
  - Packing can break during installation
  - Maldistribution of liquid

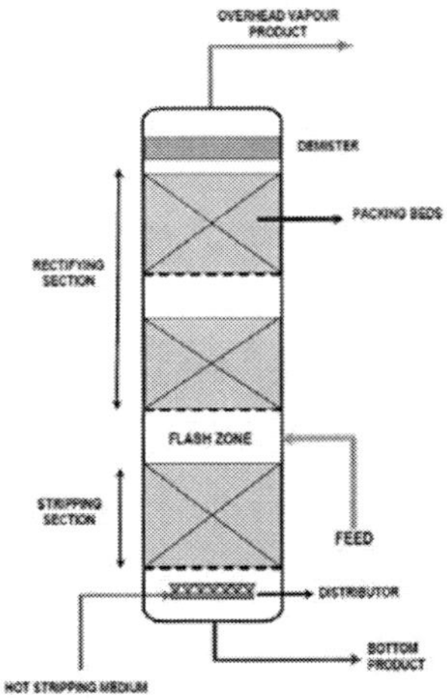

TYPES OF DISTILLATION COLUMNS

- Trayed Column
  - The number of trays is dependent on the number of equilibrium stages
- Advantages
  - Better distribution
  - Can handle high liquid flow rates
- Disadvantages
  - Higher pressure drop than packed columns
  - Foaming can occur due to induced agitation

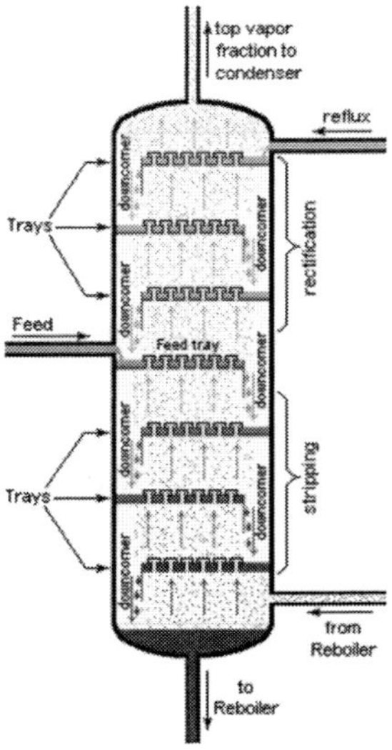

COMMON TRAY TYPES AND USES

	DUAL FLOW	SIEVE	FIXED VALVE	MOVING VALVE
CAPACITY	VERY HI	HI	HI	HI
PRESS DROP	LO	MED	MED	MED/HI
ENTRAINMENT	LO	MED	MED	MED
TURNDOWN	1.5:1	2:1	2.5:1	4-5:1
FOULING TEND.	VERY LO	LO	LO	MED/HI
RELATIVE COST	<1	1	1.1	1.2
APPLICATION	FOULING	GENERAL	GENERAL	GENERAL

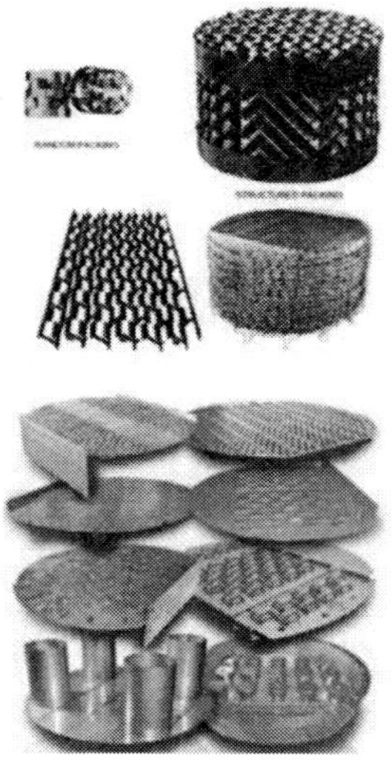

Ref.: H. Z. Kister, "Practical Distillation Technology"

BURNS McDONNELL 7

FACTORS FAVORING TRAYS VS. PACKING

TRAYS	PACKING
BETTER LIQUID DISTRIBUTION	LOWER PRESSURE DROP (~ 1/3 OF THAT OF TRAYS)
MORE PREDICTABLE PERFORMANCE	HIGHER CAPACITY IN REVAMPS
ENABLES TOWER FLEXIBILITY (MULTIPLE FEEDS, SIDE DRAWS, SIDE REBOILERS)	BETTER FOR SMALL DIAMETER APPLICATIONS
GOOD FOR CHEMICAL REACTION (DUE TO RESIDENDCE TIME)	GOOD FOR CORROSIVES (CERAMIC, PLASTIC)
BETTER FOR LOW LIQUID RATES	HAS LOWER LIQUID HOLDUP
CAN HAVE HIGHER TURNDOWN	GOOD FOR BATCH DISTILLATION (HIGHER RECOVERY)

Ref.: H. Z. Kister, "Practical Distillation Technology"

BURNS McDONNELL 8

AGENDA

- Introduction
- Types of Column Internals
- Distillation Principles
- Distillation Design: Eight Practical Steps
- Useful Resources

DISTILLATION PRINCIPLES

- A distillation column is a series of equilibrium flashes with two feeds and two product streams
- Exiting liquid is at bubble point
- Exiting vapor is at dew point
- Compositions obey the equation yi = Ki*xi

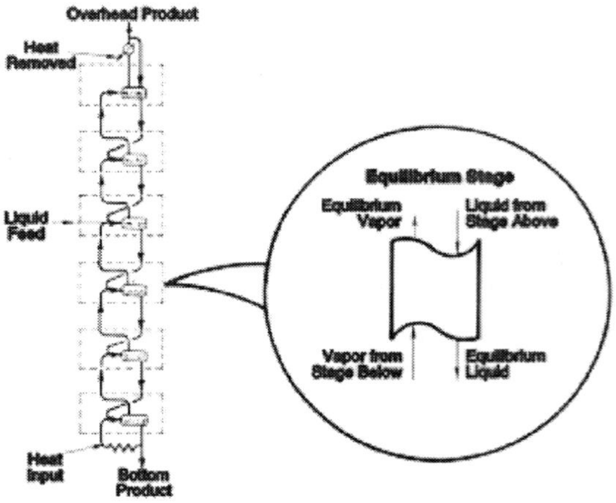

"distillation" comes from Latin "de stilla", or "of" "drop, trickle"

DISTILLATION PRINCIPLES

- A subcooled liquid ("A") is heated, its concentration remains constant until it reaches the bubble-point, when it starts to boil ("B")
- The vapor evolved during the boiling has the equilibrium composition given by "C"
- This is approximately 50% richer in component A than the original liquid
- **This difference between liquid and vapor compositions is the basis for distillation operations**

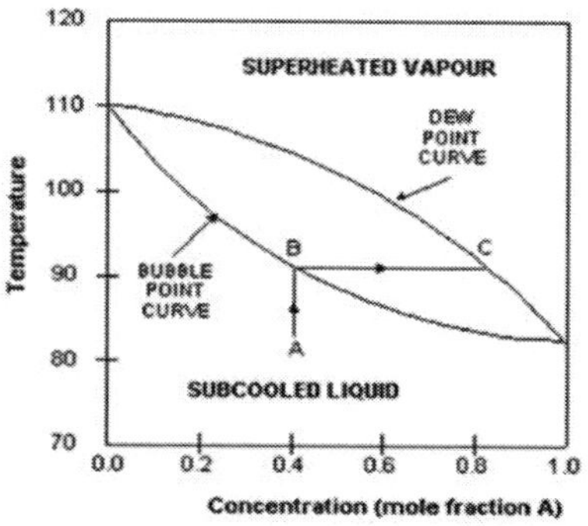

AGENDA

- Introduction
- Types of Column Internals
- Distillation Principles
- Distillation Design: Eight Practical Steps
- Useful Resources

COLUMN DESIGN: EIGHT PRACTICAL STEPS

1. Define product specification(s)
2. Choose an operating pressure
3. Choose appropriate VLE data
4. Calculate the number of theoretical trays
5. Select a tray efficiency
6. Select appropriate tower internals
7. Perform tower sizing and tray hydraulics
8. Select a process control scheme

STEP 1: DEFINE PRODUCT SPECIFICATION(S)

- Select light key and heavy key components
- There are three ways of specifying a desired product
  - A percentage recovery of a feed component in the overhead or bottom streams
  - A composition of one component in either product
  - A specific physical property, such as vapor pressure, for either product

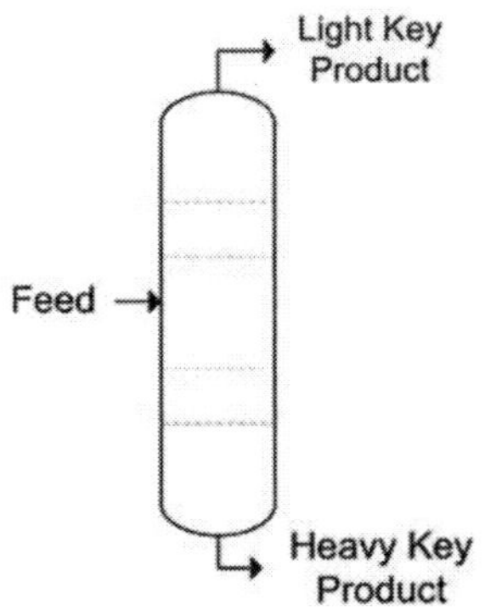

STEP 2: CHOOSE AN OPERATING PRESSURE

- Operating pressure impacts all aspects of column design
- Some considerations, as pressure increases…
  - relative volatility mostly decreases, making separation more difficult
  - the minimum number of stages increases
  - required exchanger sizes may decrease
  - The column may get mechanically more costly

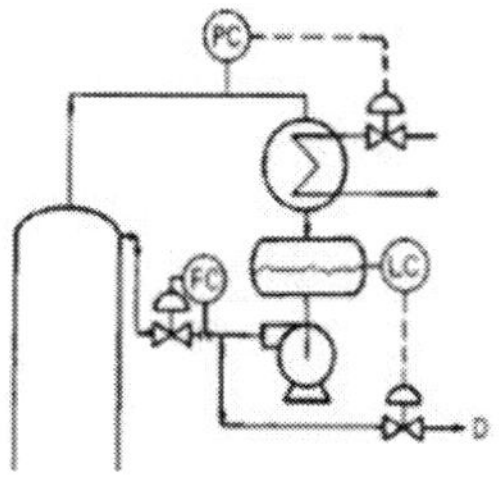

STEP 2: CHOOSE AN OPERATING PRESSURE

FIG. 19-19

Typical Fractionator Parameters

	Operating Pressure, psig	Number of Actual Trays	Reflux[1] Ratio	Reflux[2] Ratio	Tray Efficiency, %
Demethanizer	200 - 400	18-26	Top Feed	Top Feed	45 - 60
Deethanizer	375 - 450	25-35	0.9 - 2.0	0.6 - 1.0	60 - 75
Depropanizer	240 - 270	30-40	1.8 - 3.5	0.9 - 1.1	80 - 90
Debutanizer	70 - 90	25-35	1.2 - 1.5	0.8 - 0.9	85 - 95
Butane Splitter	80 - 100	60-80	6.0 - 14.0	3.0 - 3.5	90 - 100
Rich Oil Fractionator (Still)	130 - 160	20-30	1.75 - 2.0	0.35 - 0.40	Top 67 Bottom 50
Rich Oil Deethanizer	200 - 250	40	–	–	Top 25-40 Bottom 40-60
Condensate Stabilizer	100 - 400	16-24	Top Feed	Top Feed	50-75

[1]Reflux ratio relative to overhead product, mol/mol

[2]Reflux ratio relative to feed, gal./gal.

Ref.: GPSA Engineering Data Book, Chapter 19.

STEP 3: CHOOSE VLE AND ENTHALPY DATA

- Choose correct VLE and enthalpy data ranges for property calculations
- Choose correct methods for calculating properties
- Inappropriate methods, models or data ranges can lead to poor results
- Close boiling systems are especially finicky to converge
  - Watch for non-ideality!
- See your favorite simulation guru for advice

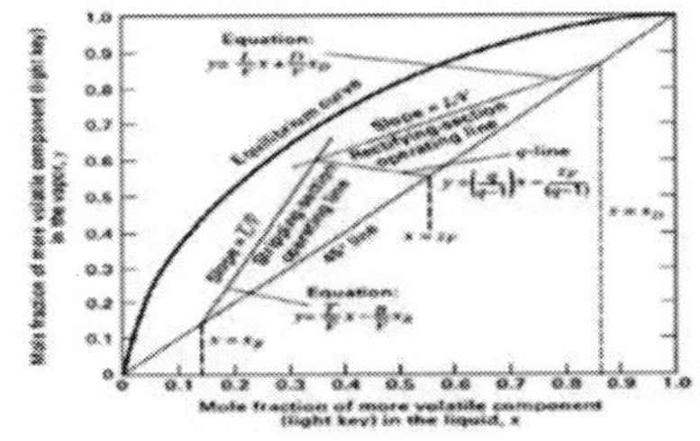

Ref.: Eric Carlson, "Don't Gamble with Physical Properties"

STEP 4: NUMBER OF THEORETICAL TRAYS

- Calculate the minimum number of theoretical trays with the Fenske equation
- For a better tray count estimate, assume a reflux ratio and run cases in a simulator
  - In Aspen Plus, use *DISTWU* model
  - In HYSYS, use *Shortcut Column* model
- All column design work is typically done with theoretical trays

$$N_{min} = \frac{\ln\left(\frac{x_D(1-x_B)}{x_B(1-x_D)}\right)}{\ln(\alpha_{AB})} - 1$$

Shortcut Method	Calculates For:
Winn	Minimum number of stages
Underwood	Minimum reflux ratio
Gilliland	Required reflux ratio for a specified number of stages or required number of stages for a specified reflux ratio

STEP 5: SELECT A TRAY EFFICIENCY

- Overall = # theoretical trays / # actual trays
- Point = (yout – yin) / (yeq – yin)
- Murphree = point efficiency, but over entire tray

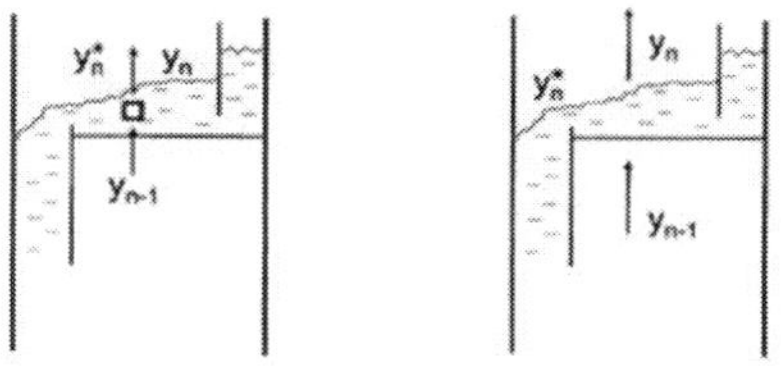

$$E = \frac{y_n - y_{n-1}}{y^* - y_{n-1}}$$

Ref.: H. Z. Kister, "Practical Distillation Technology"

STEP 5: TRAY EFFICIENCY TYPICAL VALUES

FIG. 19-19

Typical Fractionator Parameters

	Operating Pressure, psig	Number of Actual Trays	Reflux[1] Ratio	Reflux[2] Ratio	Tray Efficiency, %
Demethanizer	200 - 400	18-26	Top Feed	Top Feed	45 - 60
Deethanizer	375 - 450	25-35	0.9 - 2.0	0.6 - 1.0	60 - 75
Depropanizer	240 - 270	30-40	1.8 - 3.5	0.9 - 1.1	80 - 90
Debutanizer	70 - 90	25-35	1.2 - 1.5	0.8 - 0.9	85 - 95
Butane Splitter	80 - 100	60-80	6.0 - 14.0	3.0 - 3.5	90 - 100
Rich Oil Fractionator (Still)	130 - 160	20-30	1.75 - 2.0	0.35 - 0.40	Top 67 Bottom 50
Rich Oil Deethanizer	200 - 250	40	–	–	Top 25-40 Bottom 40-60
Condensate Stabilizer	100 - 400	16-24	Top Feed	Top Feed	50-75

[1]Reflux ratio relative to overhead product, mol/mol

[2]Reflux ratio relative to feed, gal./gal.

Ref.: GPSA Engineering Data Book, 12th ed, pg. 19-15.

STEP 5: TRAY EFFICIENCY LESSONS LEARNED

- Efficiency estimates in established processes are trouble-free with conventional internals
- With "high capacity" internals, be conservative with efficiency
- With first-of-a-kind systems, be conservative with efficiency
- There is always a learning curve with new "improved" internals --- be cautious

Ref.: H. Z. Kister, "Practical Distillation Technology"

BURNS & McDONNELL 21

STEP 6: SELECT TOWER INTERNALS

- Determine appropriate tray or packing type based on application
- Consider:
  - Fouling tendency
  - Allowable pressure drop
  - Turndown requirements
- Use Kister's table or consult vendor
- Column zoning can help capacity

	DUAL FLOW	SIEVE	FIXED VALVE	MOVING VALVE
CAPACITY	VERY HI	HI	HI	HI
PRESS DROP	LO	MED	MED	MED/HI
ENTRAINMENT	LO	MED	MED	MED
TURNDOWN	1.5:1	2:1	2.5:1	4.5:1
FOULING TEND	VERY LO	LO	LO	MED/HI
RELATIVE COST	<1	1	1.1	1.2
APPLICATION	FOULING	GENERAL	GENERAL	GENERAL

Ref.: H. Z. Kister, "Practical Distillation Technology"

BURNS & McDONNELL 22

STEP 7: CALCULATE TOWER DIAMETER

- Estimate tower diameter using rules of thumb
  - "Heat Factor" of *Q/d^2 = 350,000*
    - Duty in BTU/hr and diameter in feet
  - "General Factor" of *(R+F)/d^2 = 250*
    - Flows in BPD and diameter in feet
- Refine diameter estimate with software tools
  - KGTower
  - SULCOL

BURNS & McDONNELL

23

STEP 7: DETERMINE TRAY SPACING

- Use the table below as a guide for initial estimates
- Use KGTower or SULCOL software to rate a given internals type and tray spacing

DIAMETER	TRAY SPACING (1 PASS)	TRAY SPACING (2 OR MORE PASS)
2'-6" TO 3'-0"	18" OR 24"	----
3'-0" TO 10'-0"	24"	24"
10'-0" TO 20'-0"	30"	24"
20'-0" AND HIGHER	30" OR MORE	30" OR MORE

BURNS & McDONNELL

24

STEP 7: DETERMINE # OF TRAY PASSES

- Use the table below as a guide for initial estimates
- Use KGTower and SULCOL to rate a given number of tray passes

NUMBER OF PASSES	MECHANICAL MINIMUM DIAMETER	SUGGESTED MINIMUM DIAMETER
2	4′-6″	6′
3	7′-6″	9′
4	10′-0″	12′

TYPICAL DESIGN VALUES

- Jet and downcomer flood < 85%
- Downcomer backup of clear liquid < 40% of tray spacing plus weir height
- Downcomer exit velocity < 1.5 ft/sec
- Dry tray drop < 2" of liquid or < 15% of tray spacing
- Total tray drop < 0.1 psi per tray
- Weir load < 80 gpm/ft for one pass and < 120 gpm/ft for two or more passes
- Head loss under downcomer of 0.06" to 1.0"
- Weir height of 2" to 3"

Ref.: H. Z. Kister, "Distillation Operation"

STEP 8: CONTROL CONSIDERATIONS

- Only one of top product or bottom product composition may be controlled, unless somehow decoupled
- Composition control is readily made by temperature control unless some type of (expensive…) analysis is available
- Most columns use a material balance control scheme versus an energy balance scheme
- See Kister's "Distillation Operation", Chapters 16-19 for guidance

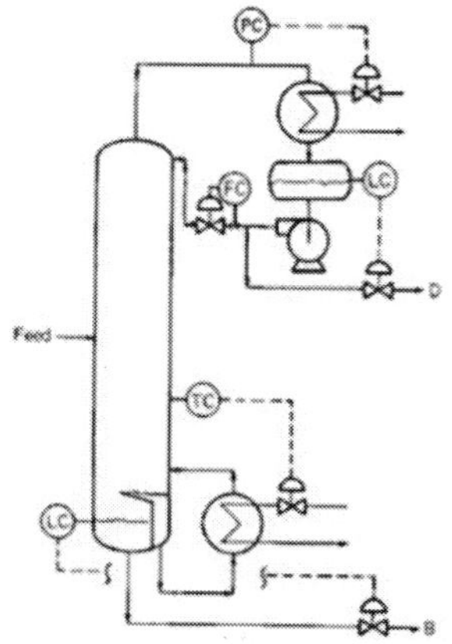

Ref.: H. Z. Kister, "Distillation Operation"

PRESSURE CONTROL

- Pressure is THE prime distillation control variable
- Affects condensation, vaporization, temperatures, compositions, volatilities… almost everything
- Pair pressure with a manipulated variable that is most 'fast acting' for good, tight control
- There are many variations on basic pressure control such as hot vapor bypass

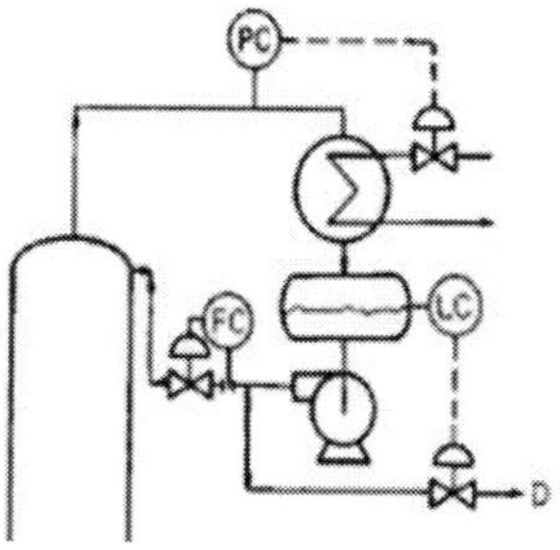

IN REVIEW: EIGHT PRACTICAL STEPS

1. Define product specification(s)
2. Choose an operating pressure
3. Choose appropriate VLE data
4. Calculate the number of theoretical trays
5. Select a tray efficiency
6. Select appropriate tower internals
7. Perform tower sizing and tray hydraulics
8. Select a process control scheme

AGENDA

- Introduction
- Types of Column Internals
- Distillation Principles
- Distillation Design: Eight Practical Steps
- Useful Resources

AGENDA

- Introduction
- Types of Column Internals
- Distillation Principles
- Distillation Design: Eight Practical Steps
- Useful Resources

BURNS & McDONNELL 31

USEFUL REFERENCES

- Kister, H.Z., Distillation Operation
- Kister, H.Z., Distillation Design
- GPSA Handbook, Chapter 19, "Fractionation"
- Branan, Carl, Rules of Thumb for Chemical Engineers, Chapter 3, 4th ed., 2005, Gulf Publishing.
- Watkins, R.N., Petroleum Refinery Distillation
- Don't Gamble with Physical Properties for Simulations, Eric Carlson, Chemical Engineering Progress, October 1996, pp. 35-46.
- Seader, J. D., Ernest J. Henley, and D. Keith. Roper. Separation Process Principles: Chemical and Biochemical Operations. 3rd ed. Hoboken, NJ: Wiley, 2011. Print.
- "ENCYCLOPEDIA OF CHEMICAL ENGINEERING EQUIPMENT." Distillation Columns. N.p., n.d. Web. 11 Feb. 2015.
- Biegler, L., Grossmann, I., Westerberg , A., 1997, Systematic Methods of Chemical Process Design, Prentice Hall.
- Bravo, Jose L. and James K. Fair. "Distillation Columns." Chemical Engineering Progress January 1990: 19-Willis, M. J., Selecting a Distillation Column Control Strategy, Department of Chemical and Process Engineering, University of Newcastle, 2000.

BURNS & McDONNELL 32

QUESTIONS AND DISCUSSION

BURNS & McDONNELL

33

DIPPR 801 - Evaluated Process Design Data

Wilding, W. V., Brigham Young University
Knotts, IV, T. A., Brigham Young University
Rowley, R. L., Brigham Young University
Giles, N., Brigham Young University
Originally presented on: 10/6/2016 13:30:00 - 13:55:00

The DIPPR 801 database, used by engineers and scientists around the world, is an evaluated database of pure-component thermophysical properties. It includes over 2000 compounds of industrial importance. Thirty-two constant properties and fifteen temperature-dependent properties are included for each compound.

The DIPPR project was initiated in 1980 as a cooperative effort between many industrial companies. A brief history of the project will be presented along with a discussion of its foundational and guiding principals.

The evaluation process used in the project seeks to ensure thermodynamic consistency among related properties. Examples of the procedures used will be presented.

DESIGN OF PIPELINE FLUSH OPERATIONS

Subramani, V., Tridiagonal Solutions Inc.
Vedapuri, D., Tridiagonal Solutions Inc.

Originally presented on: 10/6/2016 13:55:00 - 14:20:00

Pipeline flushing are routine scheduled operations in onshore, offshore oil and gas and refinery industries for various purposes, which involves liquid-liquid or gas-liquid displacement in pipelines. Some common examples are dead oil displacement of live oil in subsea pipelines, displacement of one crude by another crude in refinery transfer lines during batching process, pre-commissioning and commissioning of subsea/onshore pipelines etc. For all such operations, it is critical to accurately estimate the displacement time, circulation rate, volume of displacing fluid required, mixing length, sweeping efficiency or acceptable percentage of contamination. Industries use different approaches (correlations, analytical solutions, generic rule of thumb, CFD, commercial transient simulator, pilot scale testing/experiments etc.) to arrive at these key predictions to help in their respective flushing operations. This paper will briefly address the various approaches used in the industries. As part of this work, a few case studies were performed with a 1-D transient simulator, OLGA (licensed by Schlumberger) for illustration. Effects of flowrates and crude properties on flushing efficiency were also investigated. This paper will also identify areas for improvement in the methodologies for studying flushing.

CFD Based Thermal Fatigue Risk Assessment in Refinery Process Lines

Agrawal, M., ANSYS Inc
Venuturumilli, R.,
Khanna, S.,

Originally presented on: 10/6/2016 14:20:00 - 14:45:00

Flow turbulence at the mixing points of process fluids can generate large temperature fluctuations. If these fluctuations are transmitted to the pipe walls then they will suffer thermal fatigue induced damage over time. Often special injection devices at the mixing points mitigate the temperature fluctuations. Computational Fluid Dynamic methodology can be effectively used to predict and quantify the thermal fluctuations in the fluid as well as at the walls. An overview of the technique will be presented with the help of a few case studies.

Pollution Control and Emissions Prediction (NOx, CO, SOx) – Status Review

Jatale, A., ANSYS Inc.

Originally presented on: 10/6/2016 14:45:00 - 15:10:00

An overview of the Computational Fluid Dynamics (CFD) process as applied to full size industrial combustion processes will be presented to illustrate how CFD can be used for prediction of pollutants emission (NOx, SOx, CO etc.). An approach which couples CFD with chemical reactor network analysis for advanced pollutants emission prediction is highlighted. This approach saves on computational expense and can still predict NOx emissions from low and ultra-low NOx systems accurately. Commercial CFD package, ANSYS Fluent was used to perform all the simulations. An efficient way to quickly perform parametric study to test various pollutant control measures will also be shown.

Process Simulator Basics – Let's Build a Model!

Hill, D., Chemstations, inc.

Originally presented on: 10/6/2016 16:45:00 - 17:10:00

Do you know your way around process simulators? Perhaps you used one in college, maybe even a few times at your job's but if you're like most engineers, you probably learned how to use the program on your own through trial and error. This workshop will discuss what a process simulator does and doesn't do, and how to use its capabilities to enhance your own. Discussion will cover the basic concepts of simulation; the differences between real-world equipment and simulator models;Â decisions you need to make versus using system defaults; and best-practice advice for simulators. Portions of the workshop will be demonstrated using CHEMCAD, but the material presented applies to any process simulator's and perhaps even to models you build by hand.

Increasing Refinery Product Slate Flexibility by Revamping the Cat Feed Hydrotreater (CFHT)

Advances in Mild Hydrocracking Technology

Axens
IFP Group Technologies

Venkata Josyula
Axens North America, Inc.

AIChE
Southwest Process Technology Conference

US Distillate Market Analysis

- **5% annual increase**
- **Result of various diesel projects**

Trend Projected to Continue

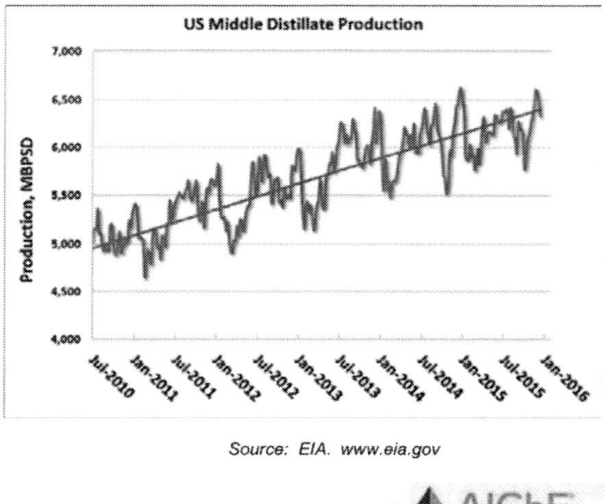

Source: EIA. www.eia.gov

AIChE
Southwest Process Technology Conference

US Distillate Export Market

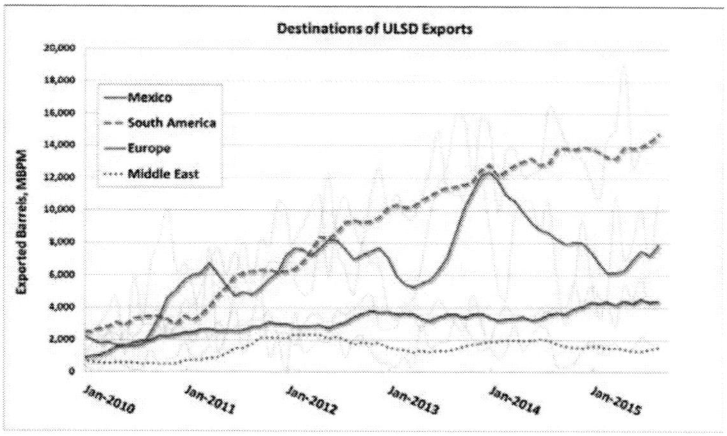

Source: EIA. www.eia.gov

Driven Mostly by Latin American Demand

World On-Road Diesel Forecast

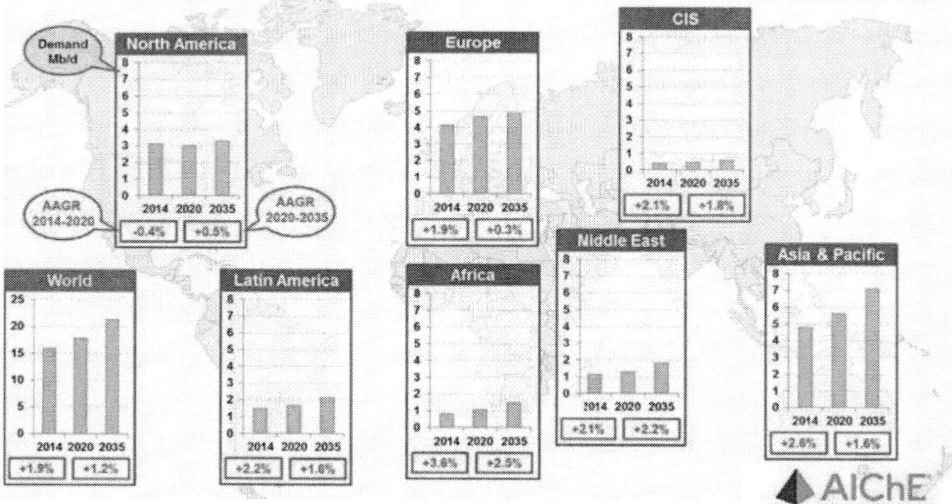

World On-Road Diesel Market

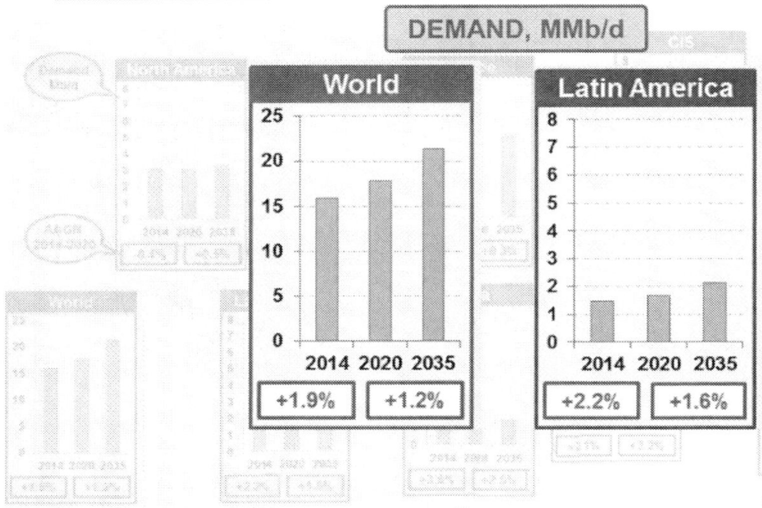

US Diesel v Gasoline Price

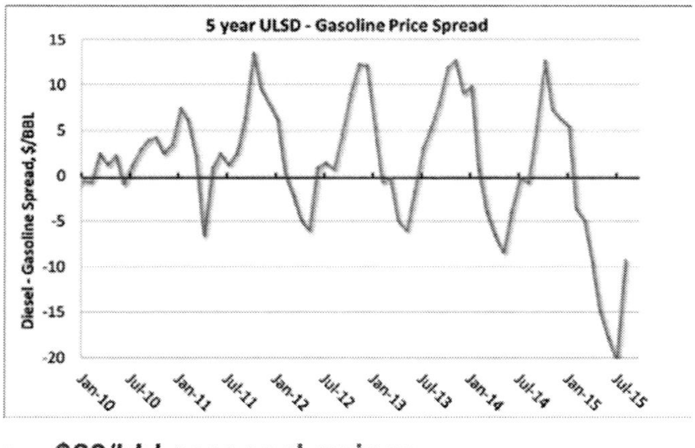

Product Slate Flexibility is Key for Profitability

- **$20/bbl** seasonal swings
- Median spread favors **diesel $2.25**
- >60% of the year, incentive to **swing to diesel**

Options for Increasing Diesel

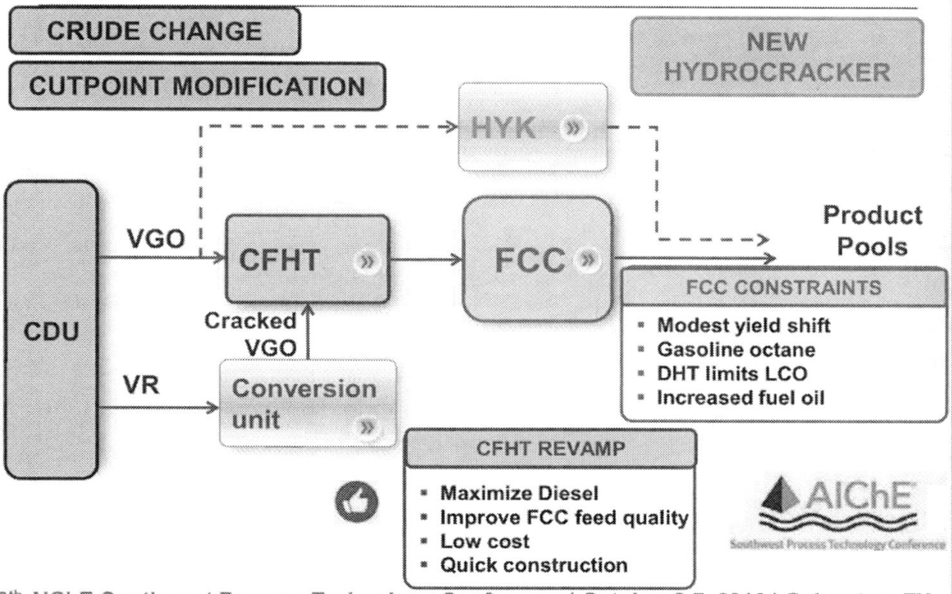

CFHT Revamp Considerations

Catalytic System → FCC Impact → Engineering Evaluation → Pool Benefits

Maximize Distillates

- Switch to hydrocracking catalyst load
- Conversion 15-35% achievable in most existing reaction systems
- Excellent selectivity towards diesel

Catalyst Solutions

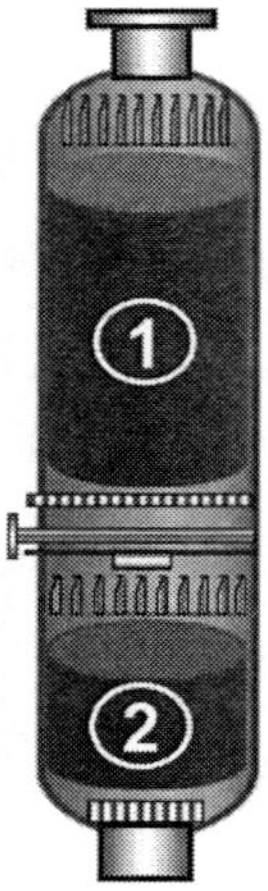

Pretreatment catalyst

Cracking catalyst

Sulfur removal

NiMo	CoMoNi
Max HDN	Max HDS

Conversion Activity

Nitrogen tolerance

HDK Series	HYK 700 Series
Amorphous	Zeolite

Tailor-made Solution Based on Objectives

Catalyst Diesel Selectivity

Axens Catalysts Developed for the Diesel Market

- R&D Geared toward European Diesel Demands
- Excellent **Diesel Selectivity**
    - Commercial results demonstrate **10:1 Diesel to Naphtha yields**
- Variety of Feedstocks (SR, CGO, DAO, LCO)
- In-line with goal of maximizing **Diesel** in the US Market

Diesel Market Driven

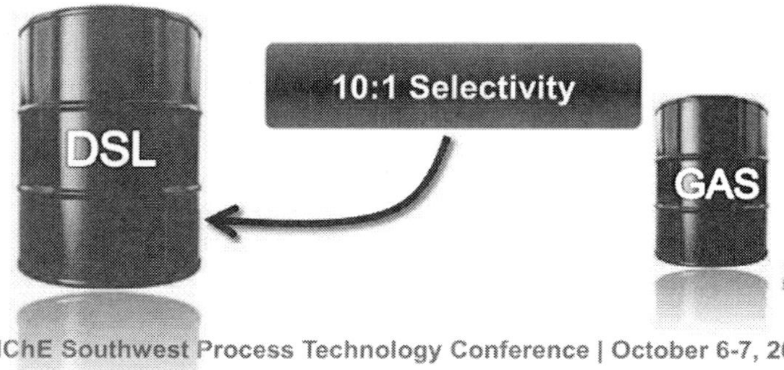

8th AIChE Southwest Process Technology Conference | October 6-7, 2016 | Galveston, TX

Revamp Considerations

Catalytic System → **FCC Impact** → Engineering Evaluation → Pool Benefits

Feed Quality & FCC Yields

- Increased FCC feed quality (aromatics saturation)
- Increased selectivity towards gasoline and LPG
- Decrease in LCO and Slurry
- Increased product quality (lower S, N)

8th AIChE Southwest Process Technology Conference | October 6-7, 2016 | Galveston, TX

FCC and Alkylation Impact

Hydrocracking increases FCC Feed Quality

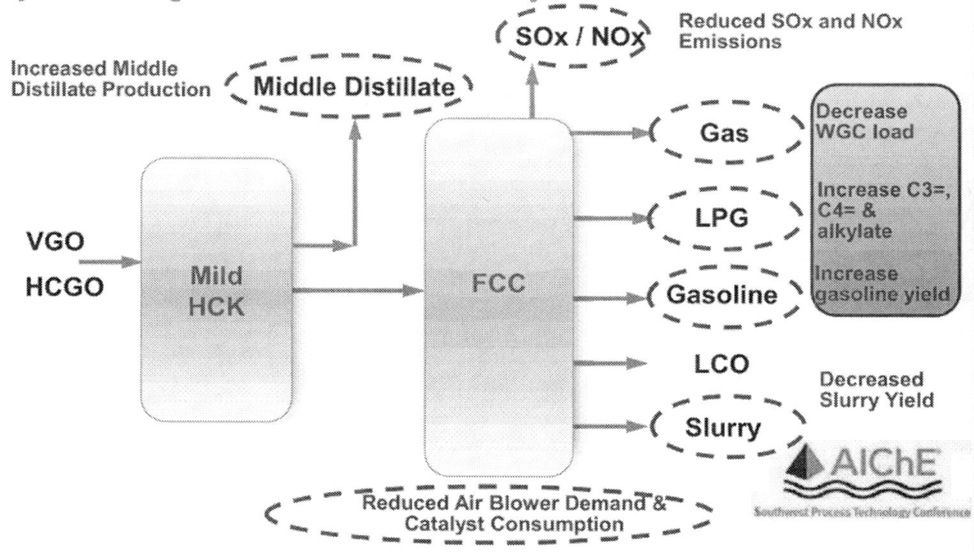

8th AIChE Southwest Process Technology Conference | October 6-7, 2016 | Galveston, TX

FCC Pretreating – H_2 Content

Hydrogen Addition Improves FCC Performance

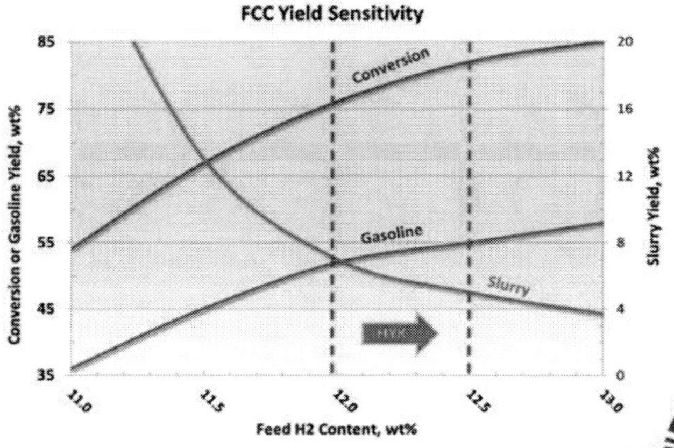

Revamp Considerations

Catalytic System → FCC Impact → **Engineering Evaluation** → Pool Benefits

Areas of Focus

- Reactor ΔP
- Reactor Internals
- Heater Limitations
- Compressors
- Stripper top Section
- Fractionation
- SAFETY SYSTEMS

Revamp Considerations

Catalytic System → FCC Impact → Engineering Evaluation → **Pool Benefits**

Improvements

- Shift D/G Ratio
- Increase overall liquid yields (volume swell)
- Reduced FCC gasoline post-treating requirements

Case Study

Southwest Process Technology Conference

Axens Case Study Results

Gasoline Mode		Diesel Mode
Refinery D/G Ratio 0.7	⟵ Operational Flexibility ⟶	Refinery D/G Ratio 1.0

› FLEXIBILITY VIA **CFHT REVAMP** TO **MILD HYDROCRACKING**:
- **Payout < 2 years**, independent of crude pricing
  - New (or Revamped) fractionator and M/U compressor
  - Modified catalyst load (Hydrocracking)
- Maintain **continuous ULSD** production
- **Maximize Octane-Barrels during summer** season

Southwest Process Technology Conference

Axens Case Study

1. **Base Case (Current Operation)**
   - Existing operation
   - FCC Feed S Target of 2,000 ppm
   - 36 month cycle in the CFHT
2. **MHC Revamp Case (Distillate Mode)**
   - Revamp CFHT to Mild Hydrocracker
   - Maximize Distillates
   - MHC Diesel treated in existing DHT unit
3. **MHC Revamp Case (Gasoline Mode)**
   - Mild Hydrocracker Revamp is complete
   - Minimize conversion & adjust cut-points
   - Maximize FCC gasoline production
4. **HyC-10 Case (Integrated Diesel HDT)**
   - Mild Hydrocracker Revamp is complete
   - Integrate new polishing reactor
   - Produce ULSD directly from the MHC

Constant Gasoline Pool RON for All Cases

AIChE Southwest Process Technology Conference

8th AIChE Southwest Process Technology Conference | October 6-7, 2016 | Galveston, TX

Case Study Configuration

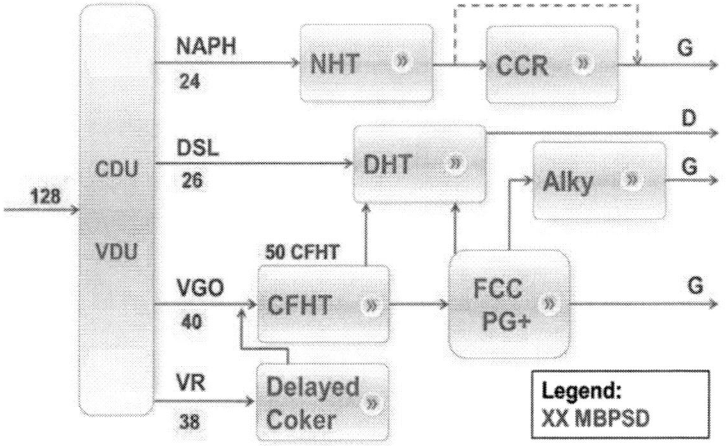

AIChE Southwest Process Technology Conference

8th AIChE Southwest Process Technology Conference | October 6-7, 2016 | Galveston, TX

Revamp – Base Case

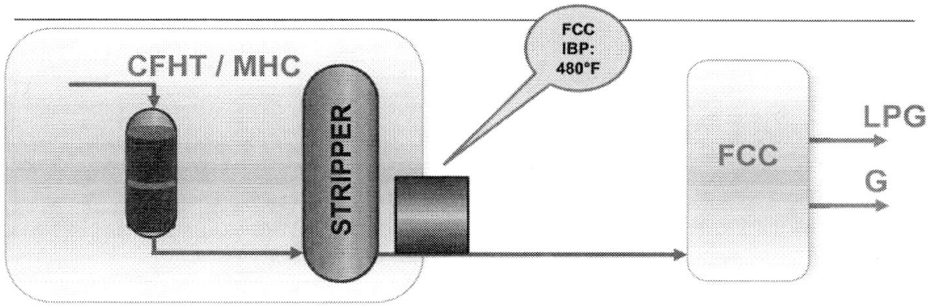

	Conversion	FCC Feed IBP	FCC Gasoline Yield
Base Case	10%	480 °F +	Base

Southwest Process Technology Conference

8th AIChE Southwest Process Technology Conference | October 6-7, 2016 | Galveston, TX

Revamp – Max Diesel

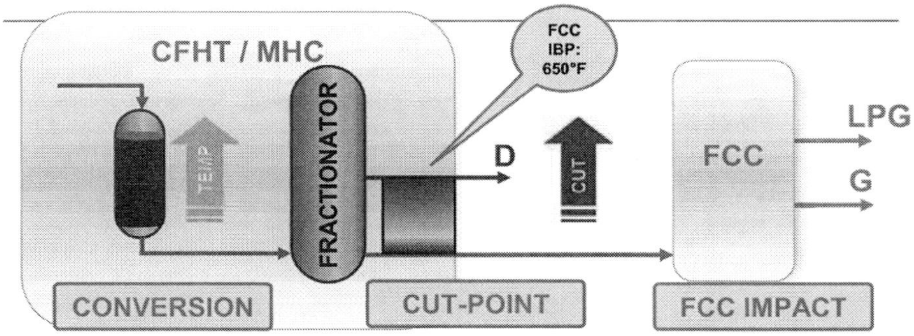

	Conversion	FCC Feed IBP	FCC Gasoline Yield
Base Case	10%	480 °F +	Base
Max Diesel	**27%**	**650 °F +**	+5.4 v%

Southwest Process Technology Conference

8th AIChE Southwest Process Technology Conference | October 6-7, 2016 | Galveston, TX

Revamp – Max Gasoline

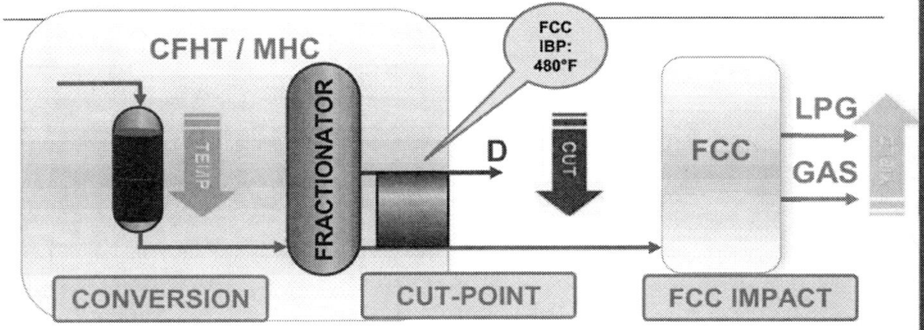

	Conversion	FCC Feed IBP	FCC Gasoline Yield
Base Case	10%	480 °F +	Base
Max Diesel	**27%**	**650 °F +**	**+5.4 v%**
Max Gas	13%	480 °F +	+1.6 v%

Southwest Process Technology Conference

Case Study Results

	Base Case	MHC Case Max Diesel	MHC Case Max Gas
CFHT/MHC			
Conversion (650 +)	10	27	13
H2 Consumption, SCFB	Base	+ 300	+ 50
Diesel Recovery, BPSD	Base	+13,050	+400

Southwest Process Technology Conference

Case Study Results

	Base Case	MHC Case Max Diesel	MHC Case Max Gas
CFHT/MHC			
Conversion (650 +)	10	27	13
H2 Consumption, SCFB	Base	+ 300	+ 50
Diesel Recovery, BPSD	Base	+13,050	+400
FCC, PG+ and Alky			
FCC Feed Rate, %	Base	-30%	Base
FCC Feed H2, wt%	Base	+0.5	+0.1
FCC Gasoline yield, wt%	Base	+5.4%	+1.6%
Gasoline Production, BPSD	Base	-8,400	+ 570
ULSD Production, BPSD	**Base**	**+9,650**	**-320**
Total Trans. Fuels Pool, BPSD	Base	+1,250	+250
Refinery Fuel Yield, v%	**92.9**	**93.9**	**93.1**

Southwest Process Technology Conference

Mild Conversion – ULSD Challenge

Mild Conversion achievable in existing units

- Shift the D:G Ratio
- Increase overall liquid yield

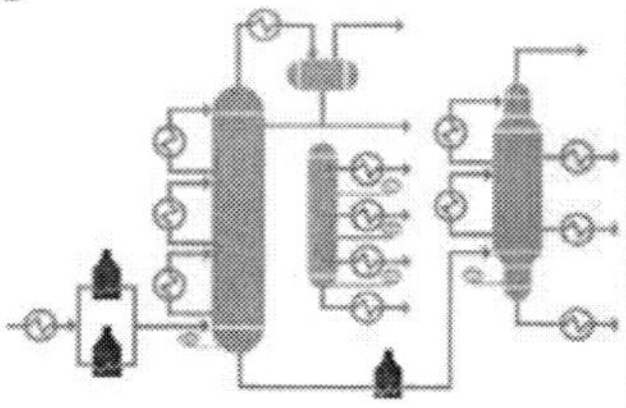

Flexibility achievable after revamp

- Return to Base Case D:G Ratio
- Extend reactor cycle

ULSD is not possible under mild conversion conditions

- DHT capacity increase is required

Axens Solution →

- HyC-10+ Process
- Integrated DHT Polishing Reactor

Southwest Process Technology Conference

Axens HyC-10 Process

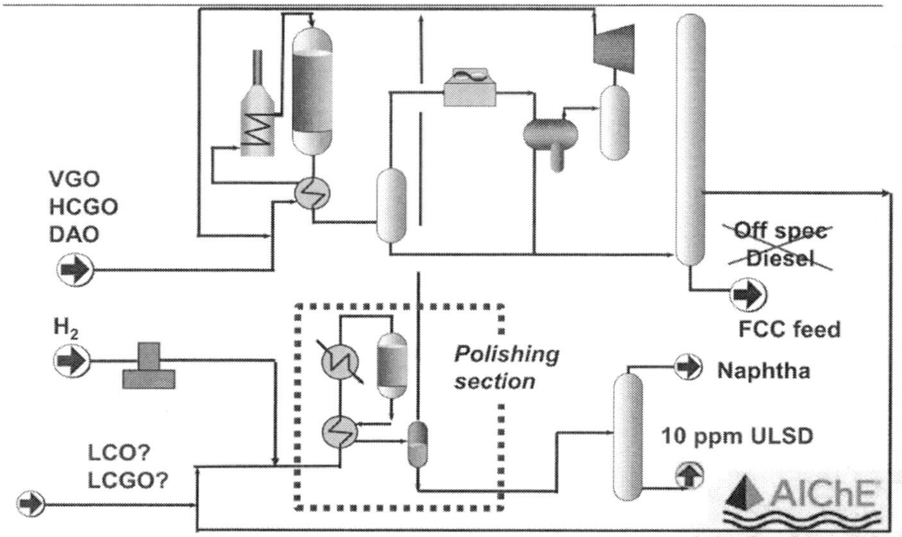

Axens HyC-10 Process

30% CAPEX Savings vs. New Unit

Polishing Reactor Operating Conditions

- Improved H_2pp due to pure make-up
- Lower reactor size due to less vapor flow and reduced H_2Spp

Major Equipment Savings vs. New DHT Unit:

- 2 compressors, 1 air cooler, HP amine absorber

Unit Integration

- Optimized H2 Usage
- Possibility for heat integration between two sections
- Maximize cycle length

8th AIChE Southwest Process Technology Conference | October 6-7, 2016 | Galveston, TX

Commercial Results

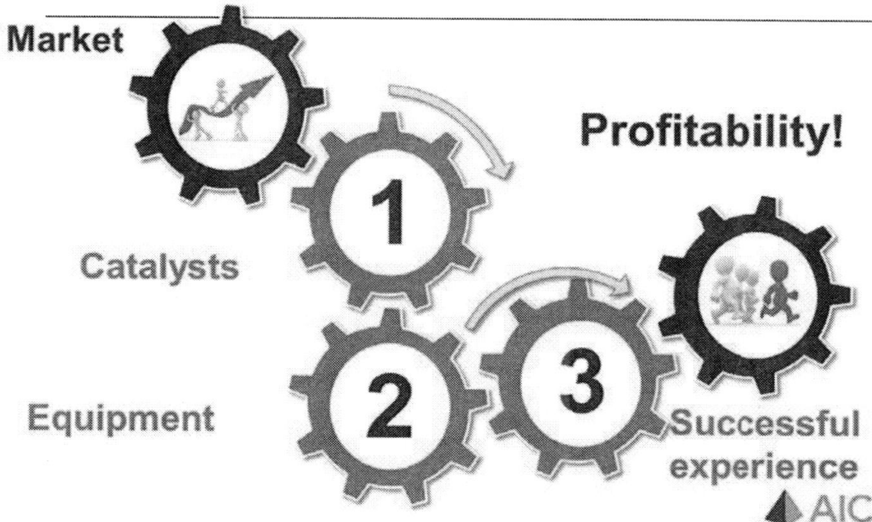

Commercial Revamp 1
Southern European Refiner

Axens Licensed Unit

- Capacity: 37,600 BPSD
- Start-up: 2005
- Catalyst loading: **100% Hydrotreating**

2010 Revamp Objectives:

- Net conversion increase
- Maximize Euro V Diesel
- Maintain feed rate
- Maintain cycle length

Axens Solution →

- **Process Study to identify bottlenecks**
- **Add EquiFlow® Internals and new catalyst combination:** HDT + Cracking Catalysts

Commercial Revamp 1
Southern European Refiner

- Revamp completed: start-up May 2013
- Feedstocks:
  - SRVGO+ Aromatic Lube Extract
  - Polishing reactor treats:
    - MHC DSL + LCO + SRAGO + VB Naphtha

Successful Revamp	Original Design	After Revamp
VGO Net Conversion, wt%	Base	+17%
Catalyst Cycle Length, months	31	On-going
MHC Euro-V Diesel Flowrate, BPSD	Base	+ 60%

AIChE Southwest Process Technology Conference

Commercial Revamp 2
Eastern European Refiner

2014 Revamp Objectives

- Increase net conversion : Maximum Diesel Production
- Maintain Product Quality
- Maintain Cycle Length
- Minimize shutdown time

Axens Solution ➡

- Process Study to identify bottlenecks
- New catalyst combination:

HR 544+ HYK 732
CoMoNi Zeolite

AIChE Southwest Process Technology Conference

Commercial Revamp 2

Eastern European Refiner

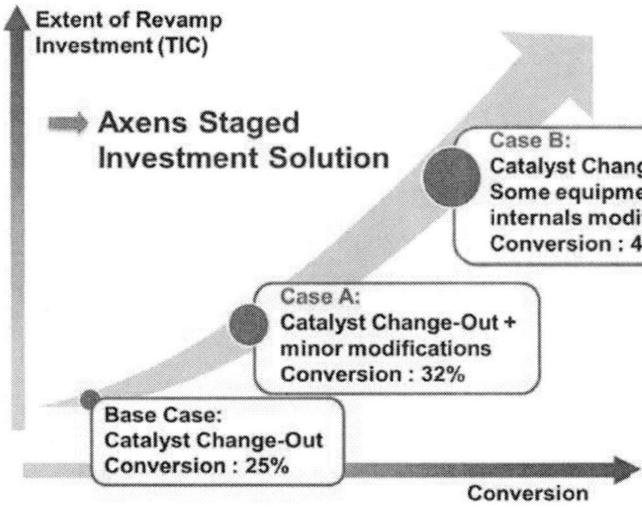

8th AIChE Southwest Process Technology Conference | October 6-7, 2016 | Galveston, TX

Revamp Economics

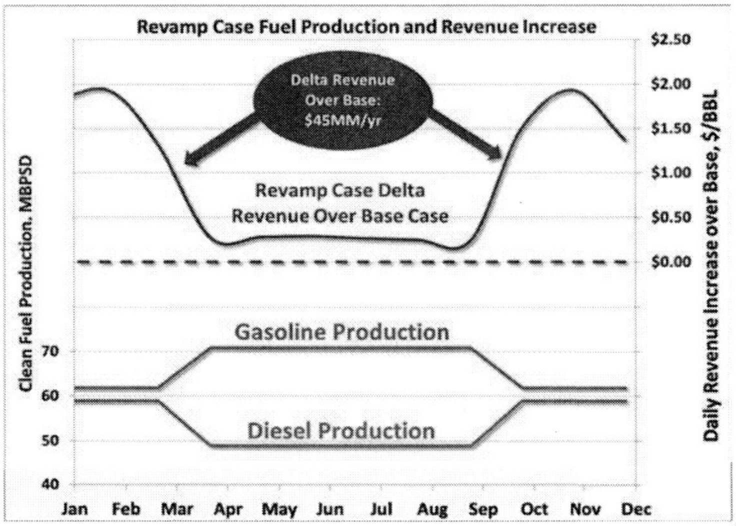

2014 Fuels Pricing. Source: EIA. www.eia.gov

8th AIChE Southwest Process Technology Conference | October 6-7, 2016 | Galveston, TX

Revamp Economics

CAPEX

- New or Modified Fractionator
- Make-Up Compressor Mods
  - **Capacity increase**
  - **Increase stroke? New machine?**
- Reactor Internals

OPEX

- Limited utility consumption increase (< 8%)
- Increased H_2 Consumption

50,000 BPSD CFHT Unit

Flexible Mode MHC

- CAPEX: $MM 30
- Increased Revenue: $MM 45/yr
- Simple Payout:

< 10 months

Flexibility → **Profitability**

AIChE Southwest Process Technology Conference

Conclusion

Revamp CFHT to Mild Hydrocracker

- **Increase diesel** production
- **Flexible** operation for high gasoline **margin** periods
- Extremely **selective catalyst**
- **Fast-track** schedule

FAST SCHEDULE **FAST PAYBACK**

HyC-10 Option

- Maintain **ULSD production** for refiners short on DHT capacity

Reduced CAPEX

AIChE Southwest Process Technology Conference

8th AIChE Southwest Process Technology Conference | October 6-7, 2016 | Galveston, TX

The New Wave of Alkylation Activity in the U.S. – Options and Economics

Lockhart, M. W., Burns & McDonnell
Stephenson, G.,

Originally presented on: 10/7/2016 8:55:00 - 9:20:00

Butanes are in cheap supply and octane value is high.Economics are driving projects towards alkylating available olefins streams as a way to bolster octane in the gasoline pool and to maximize profit.There are several alkylation projects that have been announced or are under study.These projects range from revamping existing units in refineries to constructing stand-alone grassroots facilities that are not located at a refinery, but near key NGL processing and storage facilities.This paper will review the economics driving this interest and provide an overview of the various alkylation technologies (i.e. sulfuric acid, HF, solid catalyst) and various unit configurations that are under study.Burns & McDonnell has significant experience with alkylation projects and specific examples will be given to provide a clear understanding.

New Strategies for dealing with Tramp Amines in Desalter Operation
Comer, D.,
Originally presented on: 10/7/2016 9:20:00 - 9:45:00

Tramp amines are a growing issue for the refining industry. They impact desalter performance, increase the potential for overhead corrosion, decrease neutralizer effectiveness and increase loadings on the wastewater treatment plant. Levels as low as 5 ppm of certain tramp amines can have a dramatic impact on salt points and associated corrosion in the atmospheric section.

A new class of adjunct chemistry has been identified that works in combination with emulsion breakers to mitigate the impact of tramp amines on desalter performance. The new chemistry allows the refiner greater flexibility in the selection of opportunity crudes as well as an option to potentially avoid capital investment in desalting asset upgrades. This presentation describes the innovation and presents case histories on performance in the field.

REFINED technologies® | STRATEGIC CHEMISTRY ®

The timely use of chemistry and planning designed to improve client operating margins

AIChE®

Southwest Process Technology Conference

www.r-t-i.com

REFINED technologies® | STRATEGIC CHEMISTRY ®

STRATEGIC CHEMISTRY

PATENTED CHEMISTRY + DETAILED PLANNING + PROFESSIONAL EXECUTION

- Unit Turnarounds
- <u>**Catalyst Changing Activities**</u>
- In-Situ Maintenance

2

www.r-t-i.com

REFINED technologies® | STRATEGIC CHEMISTRY®

Historical Methods for Clearing Reactors

Hot H2 Stripping

Nitrogen Purging

Today I'll talk about the power of a solvent to accelerate this process.

3 www.r-t-i.com

REFINED technologies® | STRATEGIC CHEMISTRY®

QuikTurnRX™

- Pure organic solvent (no H_2O)
- Injected into dry atmospheres (H_2, N_2, natural gas)
  - Non-water soluble; no downstream emulsions
- Residue free
- Compatible w/ catalyst
  - Catalyst can be re-used or replaced
  - Will not regenerate or poison catalyst
- Patented technology

4 www.r-t-i.com

REFINED technologies | STRATEGIC CHEMISTRY®

QuikTurnRX™ Lab Testing

- QuikTurnRX has been extensively tested by third party labs for:
  - Anions (F^-, Cl^-, NO_2^-, Br^-, etc.): <0.1 ppm
  - Cations (Na^+, K^+, Ca^+, Mg^+, etc.): <0.1 ppm
  - Metals (ICP): <1.0 μg/g
  - Other contaminants: As <1 ppb, Pb <5 ppb, N_2 <1ppm
- Distillation Range: 330°F - 380°F
- Vapor pressure: 1.4 mm Hg @ 20°C
- Full lab reports are available upon request

5 www.r-t-i.com

REFINED technologies | STRATEGIC CHEMISTRY®

Typical Process Conditions

- Chemical injected during gas sweep (cool down)
- Reactor outlet ~ 450°F and ~ 450psig
- Maximum recycle rates with specified MU and purge
- Displaced hydrocarbons routed via typical shutdown systems (flare, FGRU, slop, LP equipment, etc)
- Injection takes 2-4 hours
- Process implemented at a wide variety of conditions:
  - Reactor pressures up to 1000 psig
  - Temperatures up to 550°F; min fn of reactor conditions
  - Once through systems (no compressor)

6 www.r-t-i.com

Example Single-Reactor High Pressure Loop

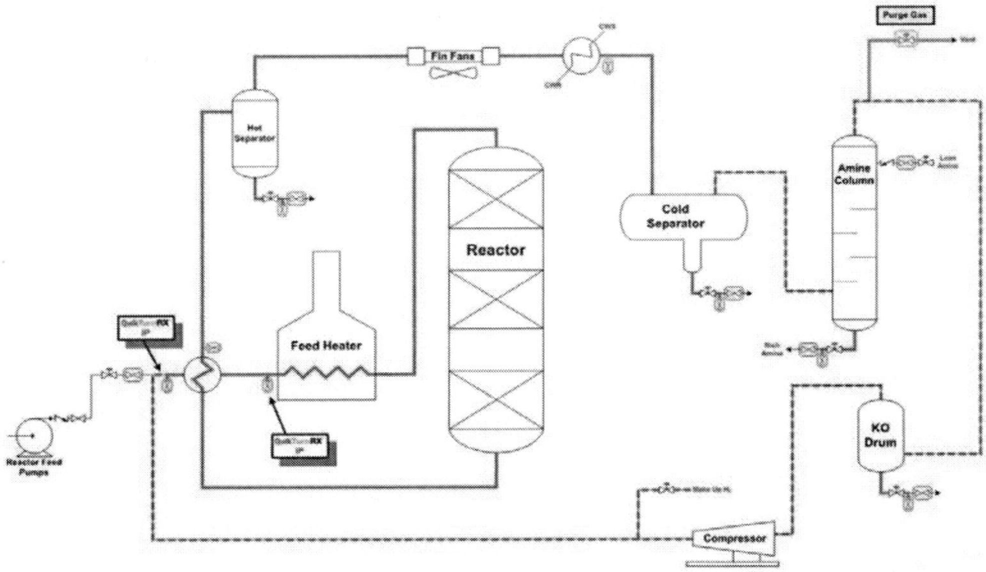

REFINED technologies® | STRATEGIC CHEMISTRY®

QuikTurnRX™ Technology Benefits

- Shortening the shutdown timeline
  - Reduced or eliminated hot H_2 sweep
  - Reduced "Huff-and-Puff" and sweeping
- No residual LEL
- Reduction in N_2 Cost
- Clean hot high pressure circuit during cool-down – maintenance cost, safety and timeline benefits
- Reduced flaring
- Predictable Results

August 2017 MACT 1 Amendment

- Deals with standards for maintenance vents that apply during startup, shutdown, maintenance, or inspection.
- Decision protocol:
  - ➢ Vessel Hazardous Air Pollutants (HAP) > 20ppm?
  - ➢ Vessel VOC content > 72#?
  - ➢ Vessel contain pyrophoric catalyst?
  - ➢ Pure H2 available at vessel?
    - ➢ If no, can purge to atm if LEL<20%. Once/yr. if LEL<35%
    - ➢ If yes, can purge to atm if LEL<10%

QuikTurnRX process assures LEL levels are met.

9 www.r-t-i.com

Results Have Been Excellent

- Reactor entries in record time
- Acceptable LEL levels with ~2 pressure/dump cycles
- Reduced or eliminated most hot hydrogen strips
- Elimination of recurring LEL and H2S issues
- Reductions in N2 usage
- No environmental, effluent or safety issues
- All hot high pressure equipment is LEL, H2S free

10 www.r-t-i.com

REFINED technologies® | STRATEGIC CHEMISTRY®

Spent Catalyst from 2006

Spent Catalyst after QTRX application 2011

11

www.r-t-i.com

Actual Gas Tests from Hydrotreater Project

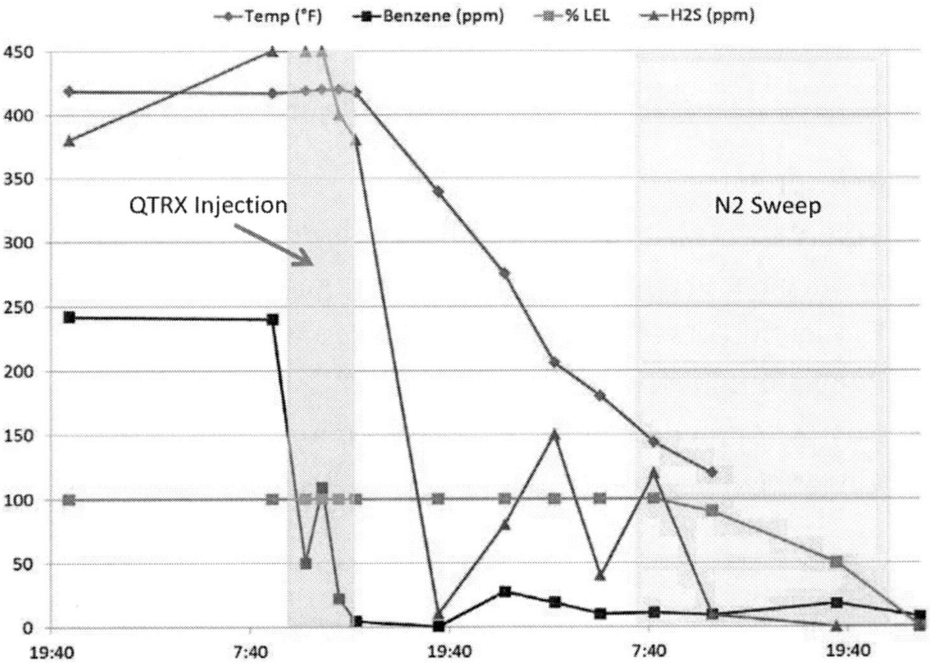

HDS Actual Time Savings

100K bbl/day HDS Unit - Previous Shutdown Timeline											
TASK	HRS	Day 1		Day 2		Day 3		Day 4		Day 5	
		Days	Nights	Days	Nights	Days	Nights	Days	Nights	Days	Nights
Stop feed, flush and prepare for hot strip	26										
Hot Strip min of 16 Hrs - or until no more liquid accumulation	16										
Cooling to 350° and lowering pressure to 1 psig	18										
Pressure dump and fill with N2 and restart compressor	10										
NiCool Injection until bed temps below 150°F and skins below 250°F	48										
Total Shutdown Duration - 118 Hours											

TASK	HRS	Day 1		Day 2		Day 3		Day 4		Day 5	
		Days	Nights	Days	Nights	Days	Nights	Days	Nights	Days	Nights
Stop feed and perform diesel flush	8										
Sweep reactor loop and cooldown to 500°F	6										
Cooldown to 450°F, pressure to 450psig and connect injection and drain piping	12										
Refined Technologies Chemical Injection	4										
Cooldown with H2 to >120°F Bed / >300°F Skin temperatures	30										
Shut down RGC, depressure system and perform two N2 pressure sump and fills	10										
Shutdown NiCool - Depressure system - Test LEL & H2S)	4										
Total Shutdown Duration - 68 Hours											

REFINED technologies® | STRATEGIC CHEMISTRY®

Learning Events

- Gulf Coast Hydrocracker
  - Unaware of quantity of polymeric anti-foulant added
  - Reactor full of long-chain polymers – goo – and sheen
  - **LESSON:** discussions on additives and quantity
- Midwest Diesel HTU
  - Plan was to use make-up PSA Hydrogen source, but used reformer due to plant issues
  - Unaware of C6+ content (~3%). LEL and benzene post cleaning.
  - **LESSON:** GC sample of make-up H2. C6+ <1%.
- Gulf Coast GOHTU
  - Project awarded a month before TA; no completed procedure.
  - Plant resistance to hooking up or draining low points
  - H2S levels stayed high – liquid in effluent train
  - **LESSON:** Must drain low points

14 www.r-t-i.com

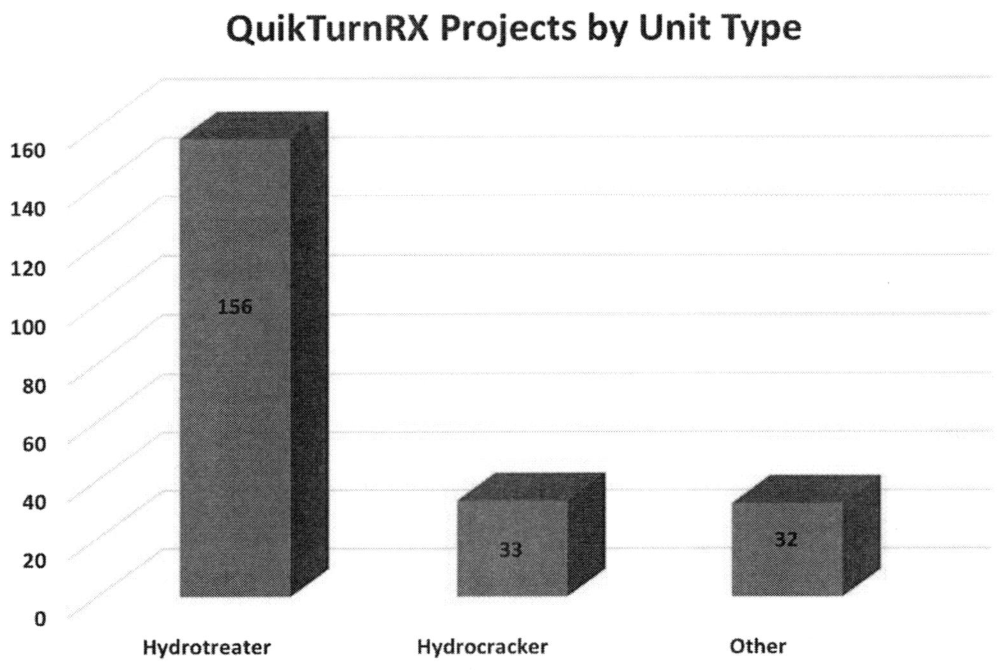
QuikTurnRX Projects by Unit Type
160
140
120
100
80
60
40
20
0
156
33
32
Hydrotreater
Hydrocracker
Other

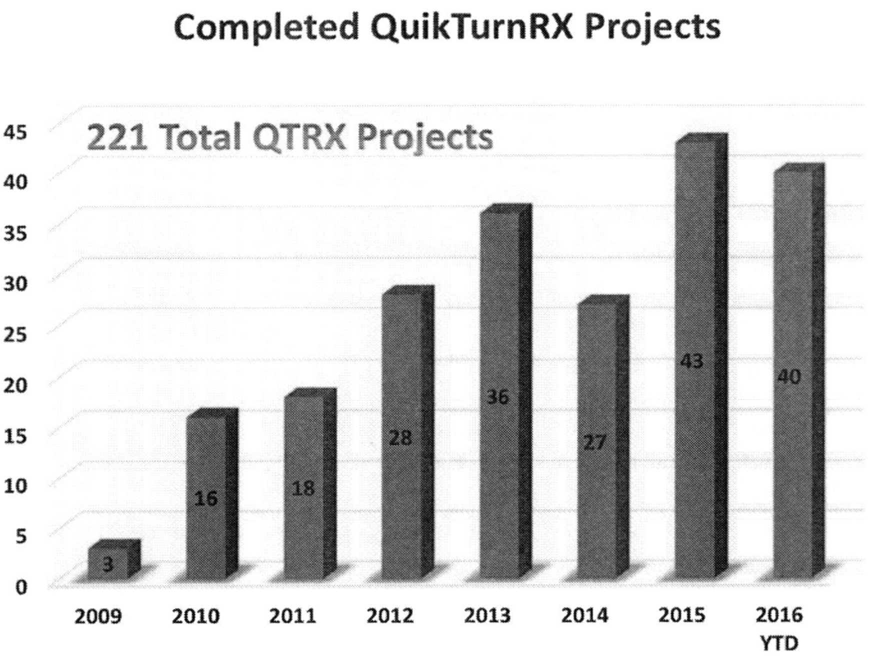
Completed QuikTurnRX Projects
221 Total QTRX Projects
45
40
35
30
25
20
15
10
5
0
3
16
18
28
36
27
43
40
2009
2010
2011
2012
2013
2014
2015
2016 YTD

REFINED technologies® | STRATEGIC CHEMISTRY®

Catalyst Compatibility Study

- QuikTurnRX was tested with DHT catalyst at a major oil refinery hydrotreating pilot plant
  - Catalyst was baselined with normal feed and hydrogen
  - Normal feed was removed from reactor system
  - QuikTurnRX was injected into the reactor with hydrogen at 450°F and 450 psig at 20X the normal commercial rate
  - After QuikTurnRX injection, normal feed was pumped back into reactor
  - Catalyst activity was measured when target rates and temperatures were achieved

17 www.r-t-i.com

REFINED technologies® | STRATEGIC CHEMISTRY®

Compatibility Study Findings

- Catalyst activity decreased slightly, but was well within the experimental error for repeatability
- The results are quite impressive considering the QuikTurnRX dosage rate
- The refiner has confidence that QuikTurnRX does not hurt catalyst activity
- Full report of pilot plant study is available upon request

18 www.r-t-i.com

Summary

- RTI QTRX process has been proven to expedite reactor entries since 2009. Record entry times achieved.
- Hot hydrogen strips and long nitrogen purges or huffs and puffs greatly reduced.
- Technology applied (221) in many unit types and sizes without any adverse effect on catalyst.
- The entire high pressure loop is hydrocarbon-free, speeding entry into piping, exchangers and other equipment.

How We Got Here

Narayan, R. S., Lyondell Chemical Company

Originally presented on: 10/7/2016 8:30:00 - 9:00:00

Feedstock costs and product values have varied with market conditions. These market drivers have influenced our choices of which chemical process technologies to apply. In addition, new technologies have been discovered.

Only 10 years ago, the perception was that the US petrochemical was on its way down with little hope of recovery. While still behind the Middle East in feedstock and energy costs, today it is a viable industry undergoing a vibrant recovery. This up and down is not a new occurrence. Dr. Narayan will explore this and set the stage for the Panel Discussion that immediately follows on Succeeding Today.

Some topics Dr. Narayan will cover include:

- Methanol synthesis (low pressure versus alternatives)
- C4 technology (Extraction versus on-purpose, dehydrogenation)
- Constrained economics situations: (coal-based in China and South Africa)
- Oil Embargo and $100/bbl oil situations versus Shale Gas and $40/bbl oil
- Major technology changes and new ones foreseen

Extending Furnace Runlength and Performance with Modeling

Nair, P., Ingenero Inc.

Originally presented on: 10/7/2016 9:30:00 - 10:00:00

The requirement for achieving the highest efficiency and maximum output from existing ethylene facilities is increasing with world-scale crackers coming online. Balancing yield, efficiency, and availability of ethylene furnaces is now much easier with the technology tools that are available. A combination of commercially available simulation software for steady state and kinetic modeling and software for empirical-based modeling can be effectively applied in combination to provide ongoing insights for improved operation. These tools are not a substitute for Advanced Process Control and Real Time Optimization but a helpful addition that can increase furnace throughput substantially while improving both yield and efficiency.

Several case studies show that achieving between 5 to 20% increase in throughput can often be achieved by utilizing readily available technology tools. Radiant section heat flux models and yield models are useful for adjusting burners to maximize run length and achieve production targets. A recent software tool in Furnace simulation Coilsim1D has been found to be extremely robust and effective. Applying Coilsim1D along with HTRI for the convection section has allowed yield maximization with run length extension. Then utilizing multivariate statistical techniques, effects on feed-stock variation can be quickly analyzed so that profitability can be maximized. The case studies to be covered discuss the technology utilized and the results found in several ethylene facilities.

Manufacturing and Automation

Tim Olsen
Emerson Process Management

Agenda

- Technical – Hands on Experience
- MBA with Business Focus
- Volunteer
- Publications

8th AIChE Southwest Process Technology Conference I October 6-7, 2016 I Galveston, TX

Tim Olsen

- 26 Years in Refining Industry
  - 18 years Emerson Process Management as Technical and Business Consultant for Refining Industry
  - 8 years UOP as Technical Advisor on Startups
- Active with AIChE
  - AIChE Fellow Member
  - Fuels and Petrochemicals Division Past Chair
  - 2014 Spring Meeting Overall Program Chair
  - Current committee member:
    - Chemical Engineering Technology Operating Council
    - Center for Energy Initiatives Advisory Board
- Active with AFPM (*prior name NPRA*)
  - Plant Automation and Decision Support Committee
- Active with Emerson Global Users Exchange
  - Board of Directors

Tim.Olsen@Emerson.com

Technical Experience

- UOP
  - Pilot Plants for 1-½ years
  - Technical Advisor and Project Manager for 4 years
  - Process Control Advisor (2 startups) and Process Control Engineering Work for 2 years
  - Data Collection and Analysis on all Refinery Units for ½ year
- Emerson
  - Refinery Consultant
  - APC, Lambda Tuning, front-end analysis and project development, economic justification

Business Experience

- MBA
  - Elective Courses included Corporate Venturing and Innovation, Developing Professional Services, Investment Banking and Financial Institutions, E-Commerce Tools, and Visual Basic programming.
- Remote Employee since 2000
  - Learn how to communicate effectively
  - Don't be invisible
- Provide strategic direction for Emerson internal
- Provide internal training on refining industry

8th AIChE Southwest Process Technology Conference | October 6-7, 2016 | Galveston, TX

Volunteer Experience

- AIChE
  - Great opportunity to volunteer and get involved
- AFPM (NPRA)
  - Much harder to volunteer as solution provider, primarily setup for operating companies – persistence pays off
- Emerson Global Users Exchange
  - Started out helping setup and teardown exhibit hall – primary reason to establish relationships and communication with Emerson's business units
  - Asked to be on the Board of Directors in 2015

8th AIChE Southwest Process Technology Conference | October 6-7, 2016 | Galveston, TX

Recent Publications

- Hydrocarbon Processing, Nov or Dec 2016, "Get more from your next turnaround"
- Offshore World, August/September 2016, "Increasing Refinery Margins with Automation and Behavior Changes"
- Process Heating, June 2016, "Improve Heat Exchanger Operation: Turn Big Data into Timely Actionable Information"
- Hydrocarbon Engineering, March 2016, "Equip for Opportunity Crudes"
- PTQ Shale Supplement, 2016, "Monitor against process problems"
- PTQ Shale Supplement 2016, Q&A, p. 5 similar to past two years
- AFPM Q&A 2015 Daily Newspaper, day 2 p. 12, "Additional measurements and timely information"
- Hydrocarbon Processing, September 2015, "Improving Refinery Flexibility and Responsiveness"
- CEP, April 2015 – "Working with Tight Oil"
- PTQ supplement on Processing Shale Feedstocks 2015 Q1, p.5
- Offshore World, Dec 2014/Jan 2015 Issue, "Smart Refinery to Enhance Productivity"
- CEP, May 2014, "An Oil Refinery Walk-Through"
- PTQ supplement on Processing Shale Feedstocks 2014 Q1, p.5,
- AFPM Annual Meeting 2014 daily newspaper, day 2 p. 14, "Shortage of skilled resources and collaboration centers"
- AFPM Q&A 2013 daily newspaper, day one cover article, "New challenges with shale oil refining and crude blending"
- CEP, January 2013, "Implement an Effective Loop Tuning Strategy",
- PTQ, January 2013, "Refinery operations gain from advances in digital automation"

AIChE Southwest Process Technology Conference

8th AIChE Southwest Process Technology Conference | October 6-7, 2016 | Galveston, TX

Questions?

Tim Olsen
Emerson Process Management

AIChE Southwest Process Technology Conference

8th AIChE Southwest Process Technology Conference I October 6-7, 2016 I Galveston, TX

A DAY IN THE LIFE OF A CHEMICAL ENGINEER – Engineering & Construction

Jaime Robles, Principal Process Engineer, Fluor Corporation
AIChE SPTC 2016 – Student Session

Best advice for ChemE's…

Regardless of where you end up, keep this in mind:

1. *Be flexible and adaptive to change.*
2. *Ask questions.*
3. *Challenge everything.*

Transitioning into the real world…

- So you've graduated (or close to)…now what?
  - Wait, what does an engineer do anyway?
    - Solve technical problems safely and efficiently to improve lives.
    - More importantly, we turn ideas into $$$$$$!
  - You are now a 'Process Engineer'…
    - Time to get paid for that brain!
    - Job searching? Starting a job? Vacation?
    - Operating, Design, Consulting, R&D

  - Life, in general…
    - Steady, more predictable schedule
    - Nostalgia for "college atmosphere"
    - Start thinking of other "grown-up' things like retirement, insurance, taxes

8th AIChE Southwest Process Technology Conference | October 6-7, 2016 | Galveston, TX

Day-to-day for Process Engineer

- Significant role throughout the course of a project
  - Go-to Team: conceptualize, design, troubleshoot
- Perform calculations…lots of calculations…
  - ✓ Process flow and optimization
  - ✓ Process simulation
  - ✓ Heat and material balances
  - ✓ Process drawings
  - ✓ Equipment specification
  - ✓ Line sizing and hydraulics
  - ✓ Instrumentation specification
  - ✓ Process safety reviews
  - ✓ Relief scenario analysis
  - ✓ Unit summaries
  - ✓ Technical support for other disciplines
  - ✓ Client interaction
  - ✓ "Data detective"

- Travel, initiatives, and other fun activities.

****Your mileage may vary****

8th AIChE Southwest Process Technology Conference | October 6-7, 2016 | Galveston, TX

"E&C Engineer At Work"

8th AIChE Southwest Process Technology Conference | October 6-7, 2016 | Galveston, TX

It starts with an idea…

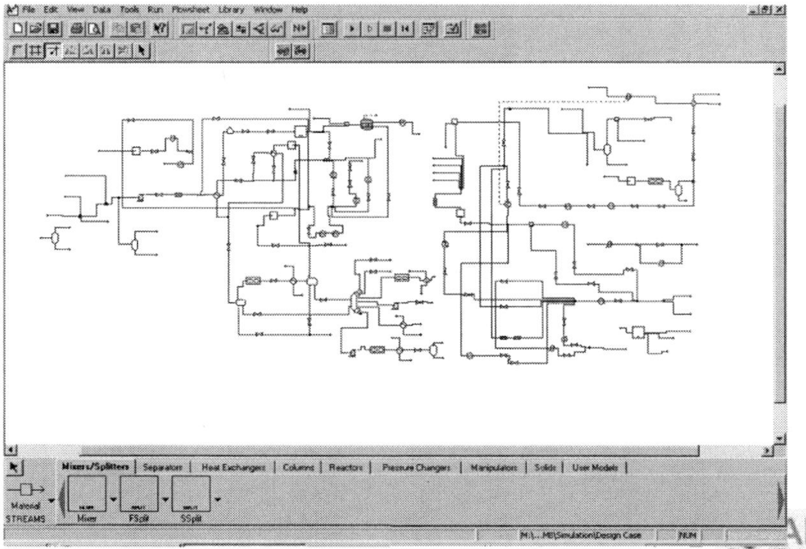

Process Simulation

8th AIChE Southwest Process Technology Conference | October 6-7, 2016 | Galveston, TX

...and it starts taking shape...

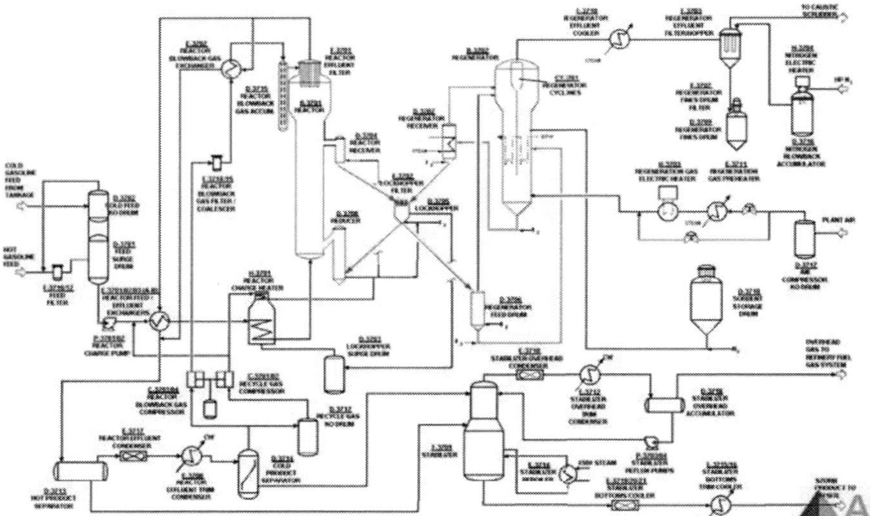

Process Flow Diagram

8th AIChE Southwest Process Technology Conference | October 6-7, 2016 | Galveston, TX

...the details are designed...

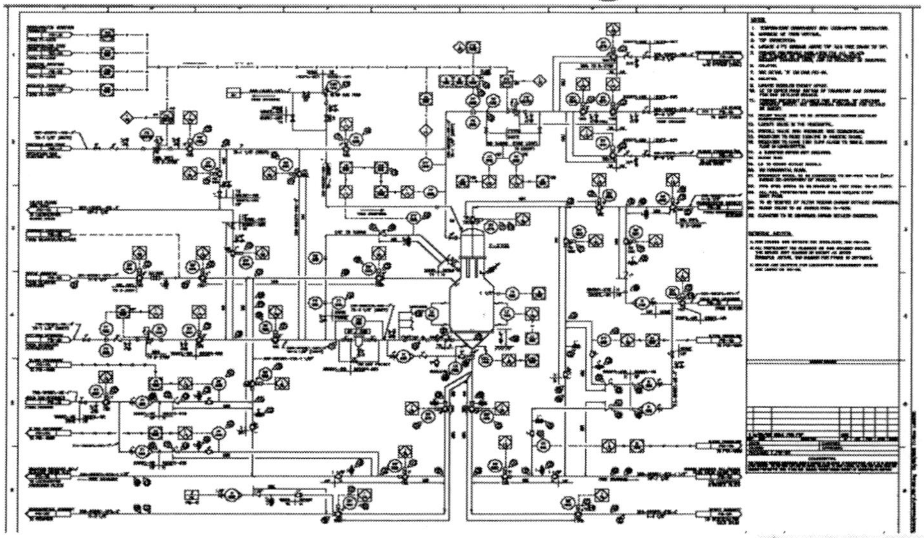

Piping & Instrumentation Diagram

8th AIChE Southwest Process Technology Conference | October 6-7, 2016 | Galveston, TX

...a few more details...

3D Model

AIChE
Southwest Process Technology Conference

8th AIChE Southwest Process Technology Conference | October 6-7, 2016 | Galveston, TX

Career paths for a ChemE's

- Honestly, your options are almost endless!

- Technical SME
- Management
- Academics
- Health
- Law
- Finance

AIChE
Southwest Process Technology Conference

But don't forget to have some fun…for your sanity!

- Stay active and involved, both inside and outside of work!

BPMS150 Fluor Cares RICE

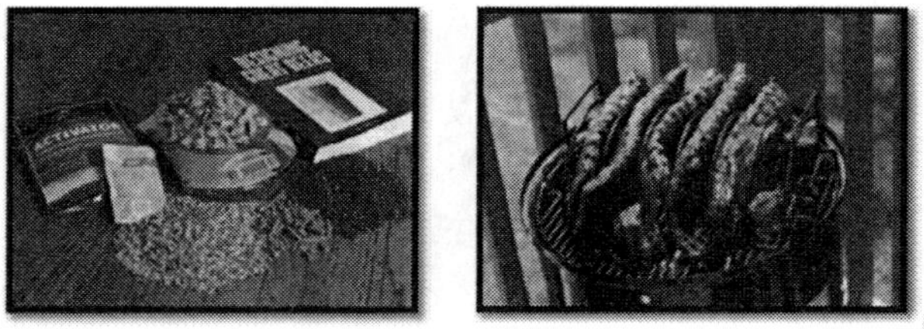

AIChE
Southwest Process Technology Conference

Survival tips

- Be organized, be disciplined, stay on task.
- Stay mindful of the future position you want.
- If you run into a problem, always offer a solution first.
- Sharpen communication skills.
- No inhibitions – ask about designs and decisions made.
- Adapt to change; be flexible; there's not just "one" answer.
- Step up to challenges...and menial tasks.
- Make your supervisors look good.
- Learn something else besides Chemical Engineering.
- See the forest **and** the trees.
- Be humble, respectful, and responsible.
- Provide **value** to your client.
- Take vacations, find a hobby, save for rainy days.
- Be a good ambassador for Process Engineers.

8th AIChE Southwest Process Technology Conference | October 6-7, 2016 | Galveston, TX

Thank you!

8th AIChE Southwest Process Technology Conference | October 6-7, 2016 | Galveston, TX

Effective Interviewing

Greg Yeo
ExxonMobil Chemical Company

AIChE
Southwest Process Technology Conference

8th AIChE Southwest Process Technology Conference I October 6-7, 2016 I Galveston, TX

Agenda

- Before the Interview
  - Preparing as if it's an important class
    - Understand the industry and company: What are they looking for?
    - Be prepared to tell your story in a relevant way
- At the Interview
    - Be authentic, relevant and motivated
- After the Interview
    - Followup, build a network, build a career

AIChE
Southwest Process Technology Conference

8th AIChE Southwest Process Technology Conference I October 6-7, 2016 I Galveston, TX

Interviewer Interaction

- What's the interviewer looking for? *The "right" balance of:*
  - **Technical Aptitude/Capability**
  - **Demonstrated Leadership**
  - **A person they can work with**
    - *Engagement, Collaboration*
    - *Enthusiasm, Interest*
    - *Flexibility*
    - *Communication Skills*
  - ***"Grit"***
- What does the interviewer have to go on?
  - **Academic record**
  - **Resume**
  - **Face to face interview**
  - *Internship (maybe?)*
  - *Campus interview (maybe?)*
  - *Phone interview (maybe?)*

"Does this person make my team better? Is this an engineer that I want on my next project?"

Prepare

- Research the company in advance: *Most candidates do not research companies!*
  - Understand the company's business
  - What they are looking for?
- Prepare an interesting resume: *Most resumes look alike!*
  - Emphasize *your* experience and skills
  - *Make it a platform to tell your story*
- Analyze yourself and ***prepare***: *Most candidates do not prepare!*
  - Know your strengths and weaknesses
  - What do they need? Do your skills fit? Do you know what you want? Are you flexible?
  - Would ***you*** hire you for this position?
- Prepare examples, stories, questions
  - This is a future colleague. Be interesting, memorable, able to tell an engaging story.
  - ***I remember engaging stories from energized, capable candidates!***

8th AIChE Southwest Process Technology Conference I October 6-7, 2016 I Galveston, TX

If You Have A Pre-interview Phone / Lunch / Dinner Meeting

- This is part of the interview process – it matters
  - Be yourself! This person is trying to get to know you.
  - Are you the kind of person they would want to work with?
- Ask lots of questions
  - Show genuine interest. Let this meeting be part of your prep
  - This person got a job – learn from them
- Be energetic, interested, professional

Before I interview a student,
I always talk to the person who took them to dinner.

At the Interview

- Dress for Success
  - If you don't know, ask or overdress. And be comfortable with it.
- Silence your phone in advance
- Use positive body language
- Be enthusiastic, engaging, and authentic.
  - Make eye contact. Show interest. And mean it.
- Create a two way conversation
  - You are interested in this interviewer and company
  - Ask questions. Take the lead when it's offered.
  - ***Be able to tell an <u>interesting story that relates to the position/company.</u>***
    - Leadership
    - Working with / connecting with people
    - Technical depth
    - Ability to stick with tough challenges

At the Interview

- Direct the conversation about your qualifications, previous accomplishments, and interest in the job
  - Highlight how previous experiences prepared you for this position
  - *If you understand the position, you can tell the right story about yourself*

- Don't be afraid to highlight areas that company may not ask about
  - Language skills, since it may be a part of the position
  - Interest in travel – ***if it's a component of the position***
  - Long term career goals that are consistent with the company
  - Your innate skills and energy:
    1. Technical excellence
    2. Leadership
    3. Team success
    4. Grit

AIChE Southwest Process Technology Conference

8th AIChE Southwest Process Technology Conference I October 6-7, 2016 I Galveston, TX

Make it a Conversation!!

- What are the primary sources of revenue?
- Ask/talk about the key products or services
- How many of your senior leaders are engineers?
- What is a typical career path for a ChE like me?
- Do you see my experience and skills as a good fit for this position?
- What opportunities are there for training, travel, continuing education?
- What kinds of things to employees get involved with outside of the plant/office?
- Is the company growing? What drives growth? Are there global operations?

AIChE Southwest Process Technology Conference

8th AIChE Southwest Process Technology Conference I October 6-7, 2016 I Galveston, TX

After the Interview

- Before you leave, ask, "What's the next step?"
  - When can you expect to hear from them?
- Thank everyone that you talk to
- Send a follow-up note or email
- Analyze your performance
  - What should I do differently next time?
- Follow up

Be Careful, Be Thoughtful, Be Professional

- Reputation and ethics matter
- Be careful with social media – LinkedIn, Facebook, Twitter
- Build bridges – don't burn them
- Learn about industries, companies, cultures interviewing experience
- Find ways to stay current

Questions?

Or conversation later today

8th AIChE Southwest Process Technology Conference I October 6-7, 2016 I Galveston, TX

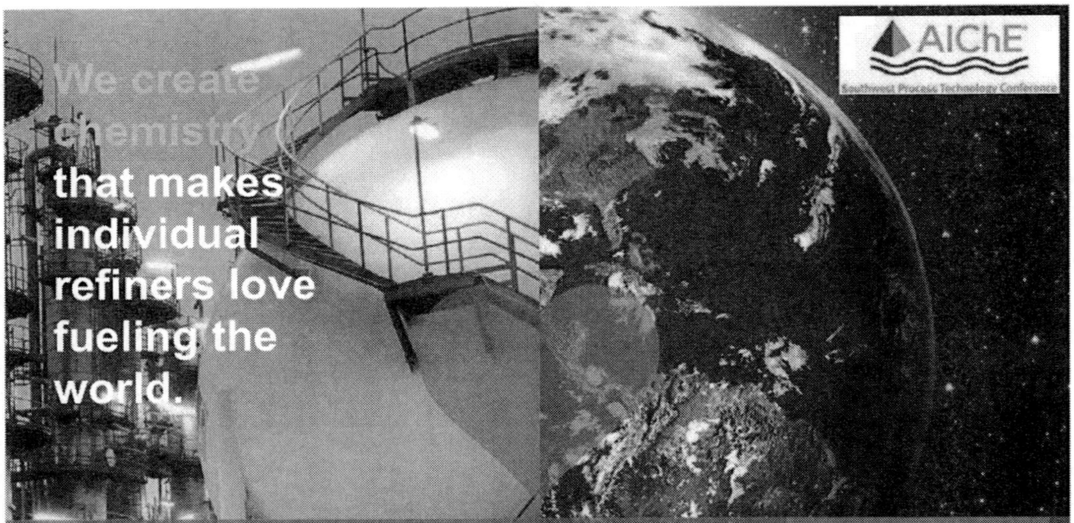

BoroCat™ – An Innovative Solution from Boron-Based Technology Platform for FCC Unit Performance Improvement

Alexis Shackleford
Technical Marketing Specialist, BASF Refining Catalysts

BASF
We create chemistry

Trends in Global Resid Operations

BASF
We create chemistry

- Ecat nickel and vanadium decreased in the US
- Ecat nickel increased significantly in the rest of the world by 37% since 2006
- The nickel challenge:
  - Dehydrogenation reactions lead to increased hydrogen and delta coke

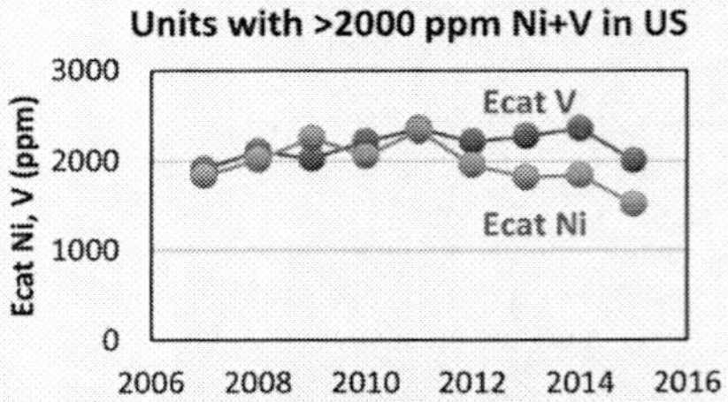

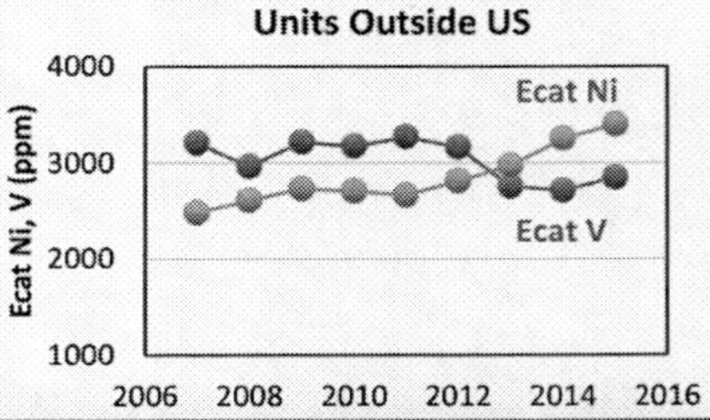

* From BASF benchmarking database. Represents 54% of all FCC units (ex. China)

Improving Dry Gas and Coke Selectivity is a Key Objective for Resid Cracking Operations

Improve Nickel Passivation for Resid Operations

BASF We create chemistry

Technology Objectives

- Create a new resid catalyst to minimize impact of contaminant Ni
- Improve conventional Ni passivation technologies
  - Specialty alumina incorporated into the catalyst
    - Immobility
  - Antimony Feed Injection
    - Environmental and safe handling concerns

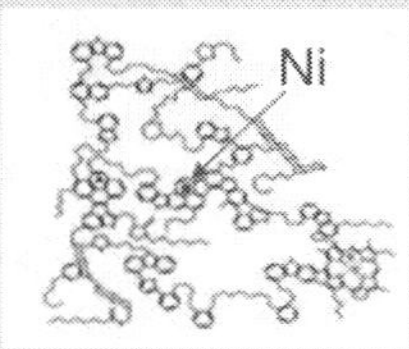

Typical metal-containing feed porphyrin

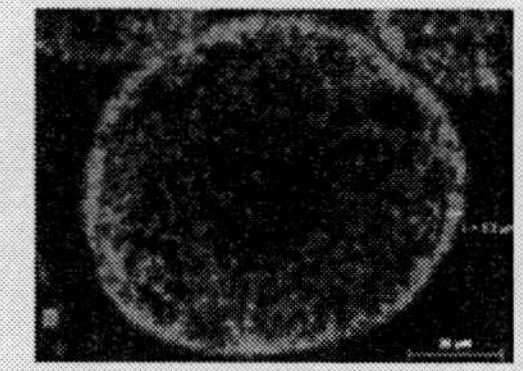

Elemental map of Ni on Ecat

A Novel Solution for FCC Refiners

BASF We create chemistry

Boron, Mobile under FCC Conditions, Migrates to Ni on the Catalyst

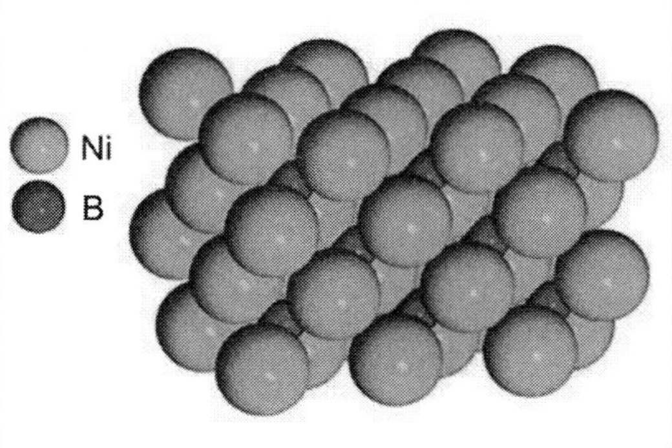

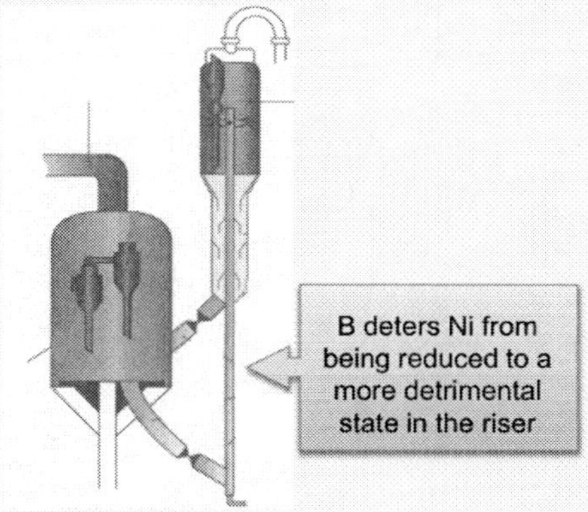

Introducing Boron-Based Technology (BBT)

BASF We create chemistry

BBT utilizes a novel chemistry for improved Ni passivation versus traditional technologies

- Boron migrates within the catalyst by solid state diffusion to passivate Ni
- Passivation of Ni confirmed by multiple spectroscopy studies

Performance Benefits

- Reduction in H_2 and delta coke

8th AIChE SPTC | October 6-7, 2016 | Galveston, TX 5

Nickel Passivation with BBT Confirmed in Multiple Lab Studies

BASF We create chemistry

BBT versus Base Resid Catalyst: Relative Yield Shift Percentage @ 75% Conversion

	CPS (3000 ppm Ni; 3000 ppm V)	CMDU (3000 ppm Ni; 3000 ppm V)	Conventional Steam Deactivation (3000 ppm Ni)
H_2 %	-27	-27	-25
Gasoline + LCO %	+0.8	+2.0	+4.0
Coke %	-22	-13	-35%

8th AIChE SPTC | October 6-7, 2016 | Galveston, TX 6

BBT Passivates Nickel without Increasing NOx

BASF
We create chemistry

Antimony use is not always possible in locations under strict NOx regulations.

Unlike antimony, BBT achieves Ni passivation without any increase in NOx

Third Party Lab Testing Results

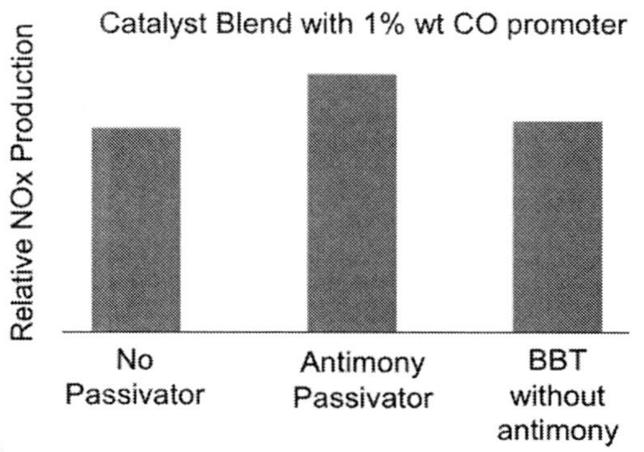

BoroCat™ – First FCC Catalyst based on BBT Platform for Resid FCC

BASF
We create chemistry

- Pore architecture that minimizes diffusional limitations of heavy feed molecules
- Reduced H_2 and delta coke
- Improved yield selectivity
- Provided operational flexibility

List of BoroCat Users with Average Feed API and Ecat Ni/V

Refinery	Location	Feed API	Ecat Ni	Ecat V
A	N America	23	2000	4000
B	N America	25	1100	2900
C	N America	31	2700	1100
D	Europe	19	3500	3200
E	Europe	23	3500	2500
F	Europe	19	2300	3500
G	Europe	21	5600	6100
H	Europe	26	2800	1600

Successful Commercialization

BoroCat Commercial FCC Trial #1

BASF
We create chemistry

- Resid unit in Europe
  - API gravity: 24 – 27
  - Ecat Ni: 2700 – 3300 ppm
  - Ecat V: 1500 – 2000 ppm
- Switched to BoroCat for max gasoline yield

Result: Improved yield selectivity with lower H_2 and delta coke

Ecat Hydrogen Reduction with BoroCat

BASF
We create chemistry

- Ecat H_2 decreased at constant equivalent Ni level with BoroCat
- Mobility of boron resulted in more effective Ni passivation

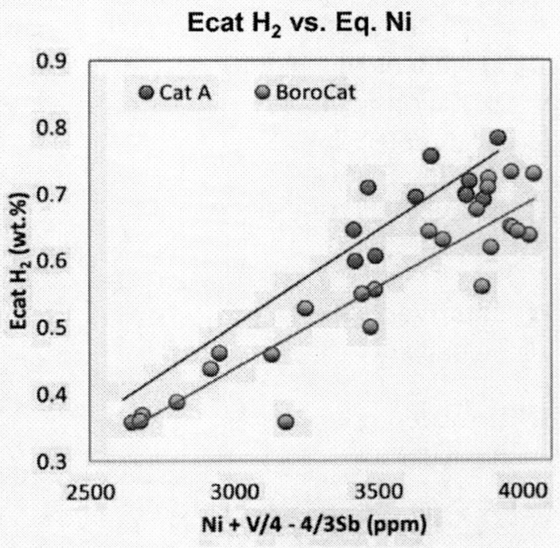

Ecat Delta Coke Reduction with BoroCat

BASF
We create chemistry

- Ecat coke factor indicates coke generating tendency of Ecat
- Coke factor decreased with BoroCat at constant equivalent Ni

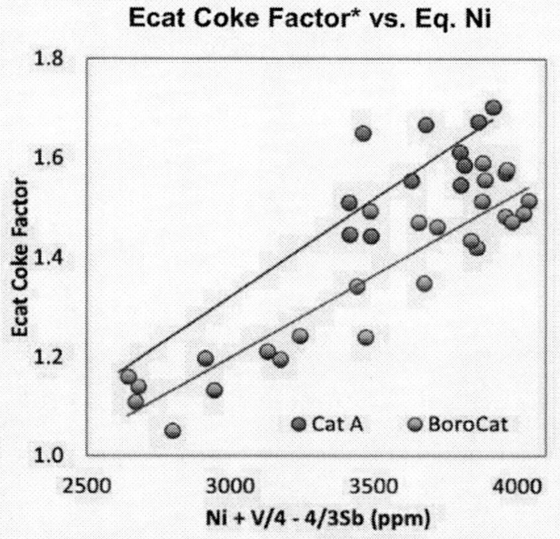

*ACE coke yield normalized by catalyst activity, a measure of coke selectivity

Improvement of Ecat Gasoline Selectivity with BoroCat

BASF
We create chemistry

- Gasoline yield increased at constant Ecat activity with BoroCat

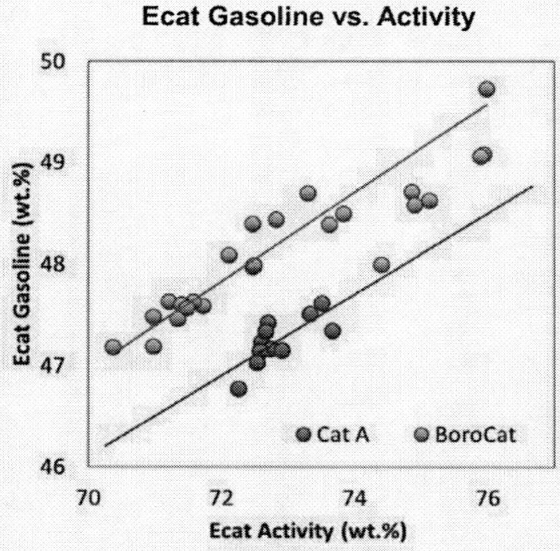

Operating Hydrogen Reduction with BoroCat

BASF We create chemistry

- Operating H_2/CH_4 decreased with BoroCat at constant equivalent Ni

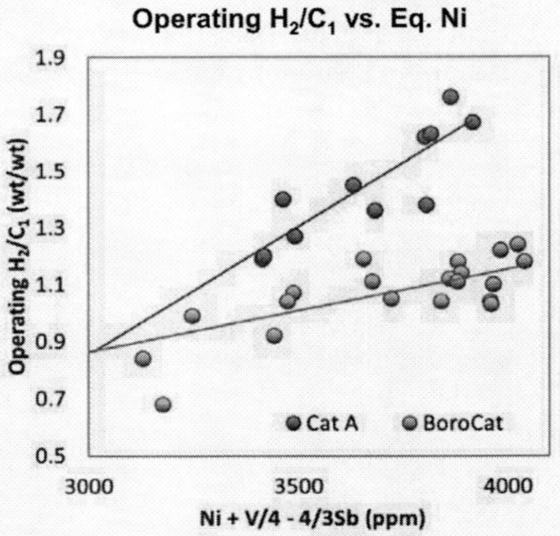

Operating Gasoline Selectivity Improvement with BoroCat

BASF We create chemistry

- Increase in operating gasoline yield with BoroCat at constant conversion

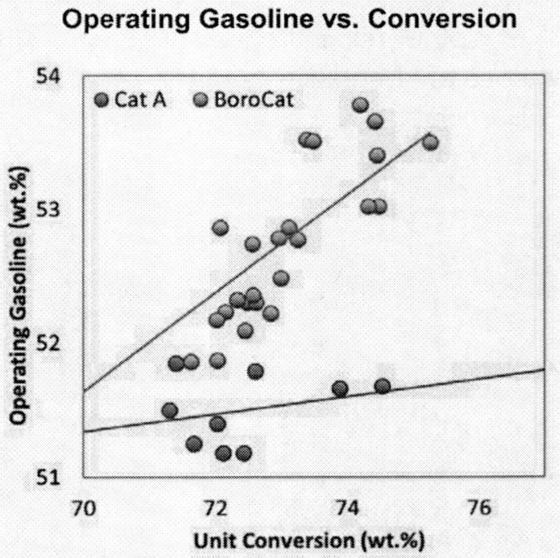

BoroCat Commercial FCC Trial #2

BASF We create chemistry

- Moderate resid unit in North America
  - API gravity: 20 - 22
  - CCR: 1 - 2%
  - Ecat Ni: 1900 – 2100 ppm
  - Ecat V: 3200 – 4000 ppm
- User of BASF Flex-Tec resid technology with antimony addition for max conversion

Result: Improved conversion with lower H_2 and delta coke

BoroCat Reduced Ecat Hydrogen

BASF We create chemistry

- Ecat H_2/C_1 ratio was lower at constant equivalent Ni
- Reduction in H_2 selectivity as expected with BBT

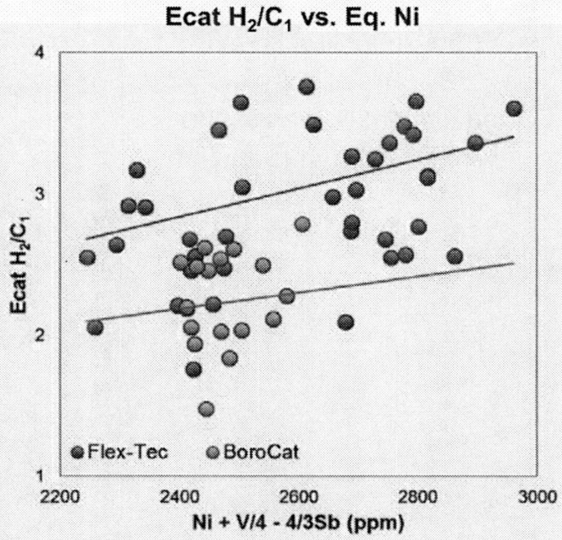

BoroCat Improved Coke Selectivity

BASF
We create chemistry

- Ecat coke factor decreased at constant equivalent Ni with BoroCat
- Improvement in coke selectivity as expected with BBT

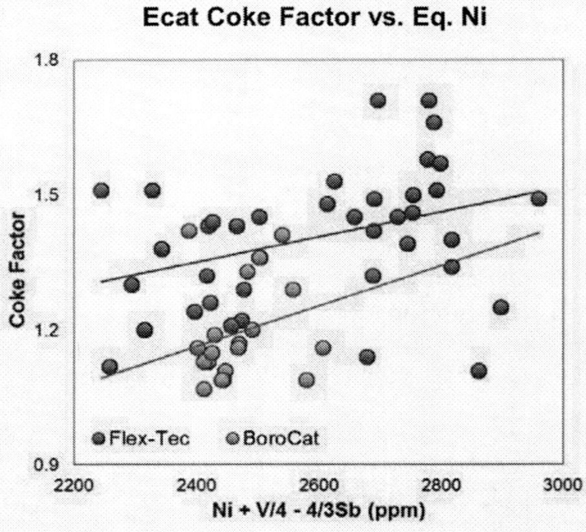

BoroCat Reduced Operating Hydrogen by ~ 25%

BASF
We create chemistry

- ~25% reduction in unit H_2/C_1 ratio at constant equivalent Ni was in line with Ecat H_2/C_1 reduction

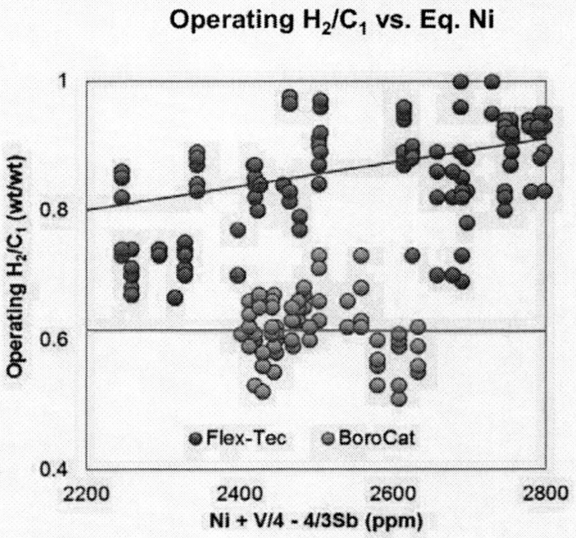

Commercial Yield Improvements with BoroCat

Profitability Improvement of $0.32-0.38/bbl

- Higher conversion, LPG and gasoline
- Lower H_2 and bottoms
- Profitability improvement in-line with refinery's internal evaluation
- Gain in operational flexibility provides opportunities to further increase profitability

At Constant Operating Conditions

	Flex-Tec	BoroCat	Delta
Conversion, vol.%	84.9	86.0	+1.1
H_2, SCF/BBL	87.1	53.4	-33.7
LPG, vol.%	27.8	28.7	+0.9
Gasoline, vol.%	52.5	53.0	+0.5
LCO, vol.%	12.1	12.0	-0.1
Bottoms, vol.%	7.7	7.4	-0.3

BoroCat Commercial FCC Trial #3

- Mild resid unit in North America
  - API gravity: 23 – 27
  - CCR: 0.4 – 1.5
  - Ecat Ni: 300 – 1400 ppm
  - Ecat V: 1300 – 3000 ppm
- User of BASF Fortress resid technology with antimony addition for Max Conversion

Result: Improved Bottoms Upgrading with No Additional NOx Emissions

BoroCat Improved Ecat and Operating Bottoms Upgrading

BASF
We create chemistry

- Higher LCO/Bottoms in Ecat data
- $0.19/bbl improvement in unit profitability

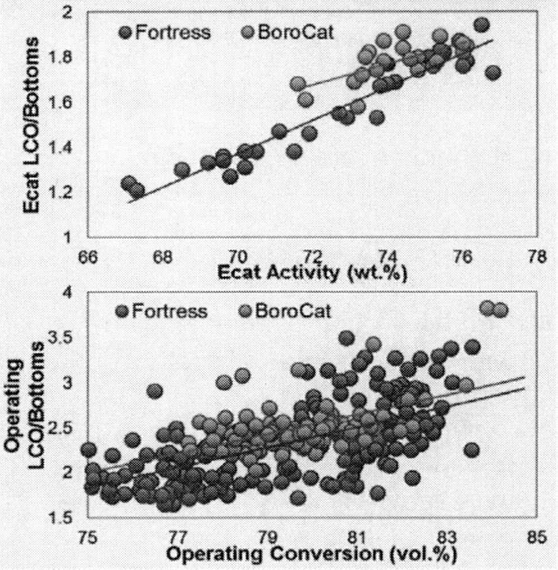

No Change in Unit NOx Emissions with BoroCat

BASF
We create chemistry

- No increase in NOx emission at constant excess oxygen

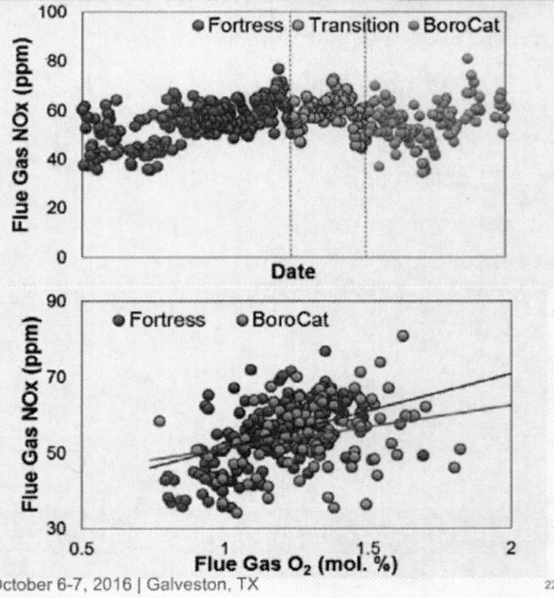

No Change in Unit NOx Emissions with BoroCat

BASF
We create chemistry

- No increase in NOx emission at constant excess oxygen

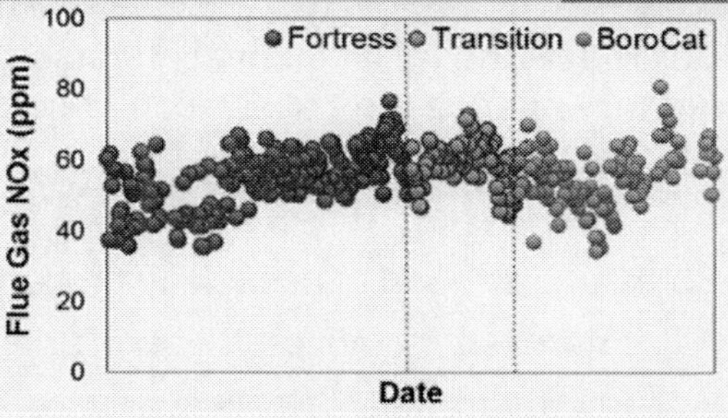

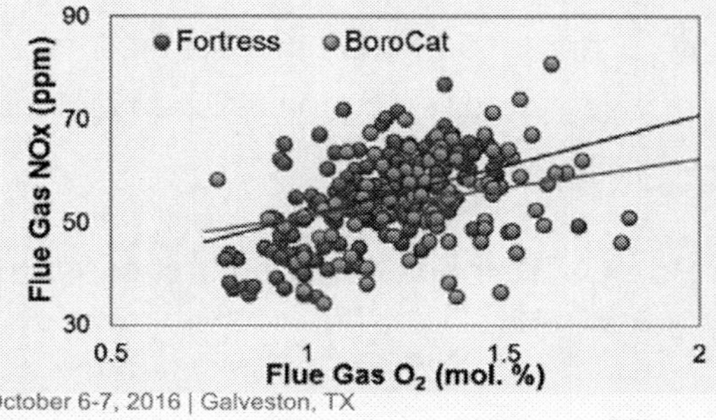

8th AIChE SPTC | October 6-7, 2016 | Galveston, TX

23

BoroCat Trial #4 Tamoil Collombey

BASF
We create chemistry

- R2R Design
- 100% Atmospheric Resid
- 21 API gravity
- 4-6 wt% Conradson
- Ni + V ranges 5,000-10,000ppm
- Uses Sb and ZSM-5

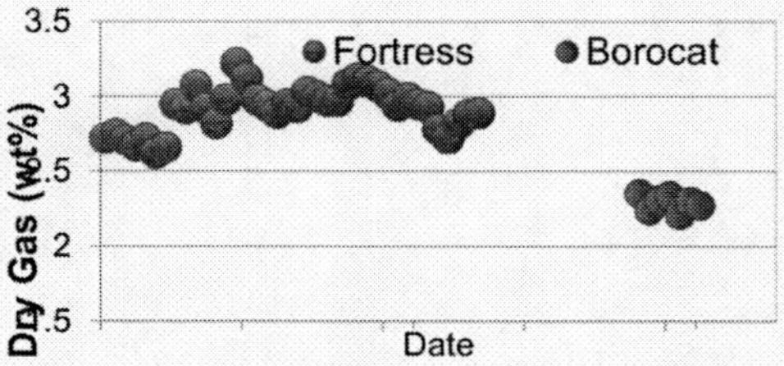

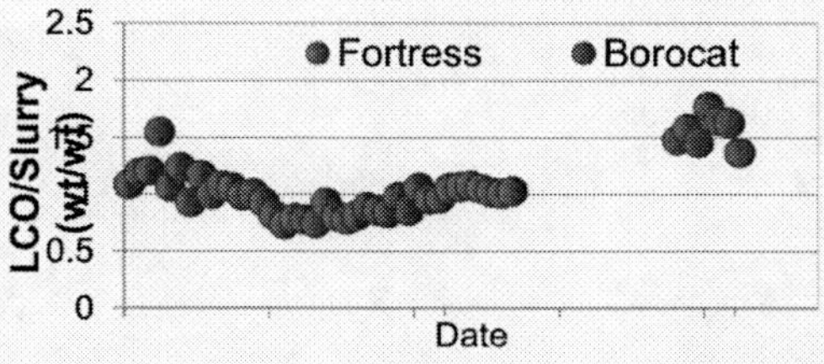

Conclusions and Path Forward

BASF
We create chemistry

- BoroCat is the latest innovation based on BBT platform designed to further enhance the passivation of contaminant Ni
- BoroCat has been successfully proven in multiple refineries:
  - Improved metals passivation and lowered hydrogen and delta coke
  - Improved yield selectivity
  - Increased FCC unit profitability
  - No increase in unit NOx emission
- BoroCat unloads common unit constraints, giving flexibility to optimize unit operations.

8th AIChE SPTC | October 6-7, 2016 | Galveston, TX 25

AIChE
Southwest Process Technology Conference

Thank you

BASF
We create chemistry

Hydroprocessing Design and Engineering Best Practices

Turini, K., Fluor Corporation

Originally presented on: 10/7/2016 13:55:00 - 14:20:00

Refinery design of hydroprocessing units has become more complex with the ever increasing demand for clean fuels. The best practices for the design and revamp of exiting units will be covered. This covers an overview of the major issues and design criteria that should be followed.

Refiner Benefits from Advanced Distillation Systems

Bhargava, M.,

Originally presented on: 10/7/2016 14:20:00 - 14:45:00

This presentation gives a case study where a global refining company implemented Dividing Wall Column (DWC) technology to improve feedstock quality in an aromatics complex, while simultaneously reducing production cost. The application demonstrates that significant cost savings can be achieved without sacrificing flexibility or performance.

DWC technology is useful to separate a multi-component feed into three or more purified streams within a single tower, thereby eliminating the need for a second tower. There is a thermodynamic efficiency gained by designing a multi-component system with pre-fractionation followed by narrow-cut fractionation, which is done by segregating zones within the distillation column. Other variants of DWC distillation are used to gain heat recovery, or accomplish different unit operations within the same equipment. Collectively, these are termed as Advanced Distillation systems.

The presentation is jointly prepared by licensor and operator, and includes the following topics:

- Basic concepts of dividing wall columns
- Case study of DWC retrofit into an existing refinery column
- Other applications where Advanced Distillation is applicable

Reducing Energy Consumption in Separations

John G. Pendergast, Dow Chemical
Senior Fellow, Oil & Gas, Mining and Water
Formerly: Fellow Process Separations, Engineering Sciences
Ravi Dixit
Fellow: Reaction Engineering, Engineering Sciences

Agenda

- Company Introduction
  - Limitations on Information
- Historical Perspective
- Economic Basis
- Energy Conservation Separation Methods
- Success Examples
- On the Horizon
- Questions

8th AIChE Southwest Process Technology Conference I October 6-7, 2016 I Galveston, TX

The Dow Chemical Company

Diversified chemical company, harnessing the power of science and technology to improve living daily

1897	2000	2008	2016
	Acquired Union Carbide	Acquired Rohm & Haas	Immediate Future Dow Corning DuPont

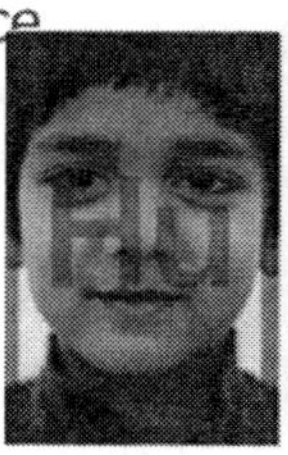

Unique perspective of commodity landscape and specialty chemicals
Founded on brine chemistry 119 years ago

Dow by the numbers

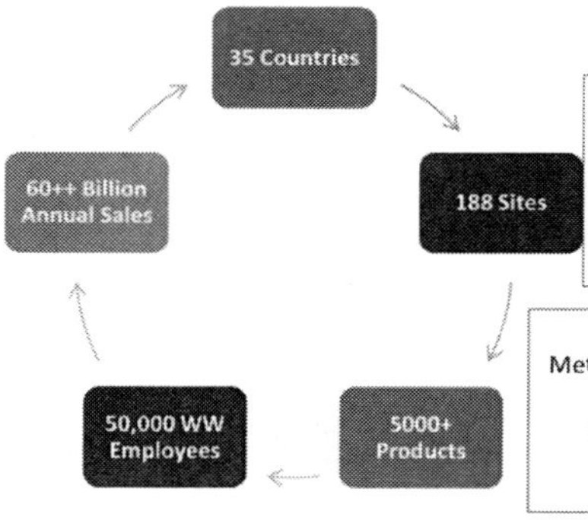

Large Scale
Ethylene and polyethylene
Ethylene oxide and ethylene glycol
Polyurethanes & Polyols
Chlorine and caustic

Specialty Scale
Metal organics, tri-methyl gallium, tri-methyl indium
OLED materials
Chemical mechanical planarization products, photoresists...

AIChE Southwest Process Technology Conference

8th AIChE Southwest Process Technology Conference I October 6-7, 2016 I Galveston, TX

Nature Magazine, April 2016,
D. Scholl, R. Lively

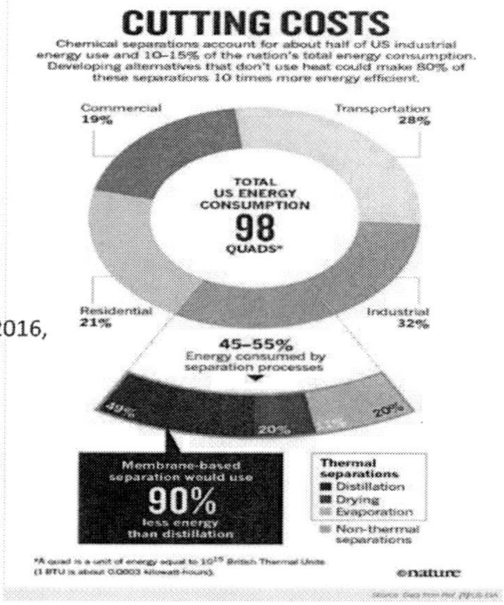

Recent Perspective

AIChE Southwest Process Technology Conference

7th AIChE Southwest Process Technology Conference | October 1-2, 2015 | Galveston, TX

The Nature of Distillation / Traditional Separations Research

- ❑Most separation technology (distillation) is viewed as mature
- ❑Many improvements will probably be incremental
  - ❑This <u>does not</u> say unimportant or trivial
  - ❑This <u>does not</u> say "all" will be incremental
    - ❑ Make no mistake ... breakthroughs will be difficult!
- ❑Industrial distillation research is largely out of favor in academia
  - ❑There are exceptions
  - ❑Viewed as mature
  - ❑Breakthroughs will require collaboration of diverse parties

AIChE Southwest Process Technology Conference

Distillation Device Development Energy Reduction

- Device Development
  - Main stream device development by suppliers
    - Specialty devices by in-house expertise as required by business needs
  - External research must be viewed as a source of potentially attractive new devices
- Device testing and validation by
    - In-house testing as needed
    - Consortia (FRI, SRP ...)
    - Ad hoc alliances
    - External research partnerships

AIChE Southwest Process Technology Conference

8th AIChE Southwest Process Technology Conference I October 6-7, 2016 I Galveston, TX

The Capital / Energy Conundrum

- Energy improvements alone will generally not support replacement investment
- Distillation improvements must support the cost of capital
  - A process may be very energy or raw material efficient, but be substantially higher in capital
  - Example: a distillation tower with a heat pump may consume less energy, but be substantially more capital intensive
- Capital is limited, and energy improvements <u>must</u> compete for capital
- Hierarchy of financial validity
  - Capacity
  - Quality
  - Energy

Subtle Aspects of the Conundrum

- Our processes (physical facilities) have <u>long</u> life spans
  - 30 years is not uncommon ... 50 years not unknown
    - Maintenance is performed and parts replaced
    - Instrumentation upgraded
    - Plants are by no means neglected .... <u>but</u>
  - Basic configuration may not have changed ... in two generations!
- How do we find and implement process alternatives to <u>improve performance</u> that meet our economic criterion
  - Energy cost alone <u>will not</u> support total replacement
    - If your organization sees this differently, great
  - Additional capacity <u>generally will</u> support capital expenditure
    - It must fit into the existing framework
    - We must explore "all" alternatives

The Nature of Separations Selection

- ❑Sequential, single feed towers are still the first choice
- ❑Integrated sequences (side-rectifiers, side-strippers ... etc.) make more efficient use of mass-transfer and heat integration opportunities with increased capital <u>investment</u>
- ❑Numerical capabilities are now available which make rigorous sequence evaluation and selection a reality
- ❑Risk aversion is real and <u>entirely understandable</u>
  - ❑Hurdle: Seen as an increased risk by Manufacturing ... flexibility and control
  - ❑Start-up and control <u>*can be*</u> more difficult than conventional sequences

Energy Saving Configurations

- ❑Complex Distillation Column Arrangements
  - ❑Petlyuk Column
  - ❑Complex Column Configurations
  - ❑Dividing Wall Columns
- ❑Hybrid Separations
  - ❑Distillation / PSA
  - ❑Distillation / membranes
- ❑Reactive Distillation
- ❑Horizons
- ❑Not Discussing: Improvements in Distillation Control
  - ❑ Distillation control offers major potential
  - ❑ Low to zero capital

8th AIChE Southwest Process Technology Conference I October 6-7, 2016 I Galveston, TX

Separation Sequencing

- We have the tools and the techniques to
    - Generate <u>all</u> sequences
    - Simple <u>plus</u> complex
    - Heat integrated sequences
  - Use optimization tools to decide the "best" sequence
- Our challenge
  - Use these tools to build new plants
    - Even more challenging: Use these tools to find better sequences that fit our <u>existing</u> facilities

Extremely challenging but very important

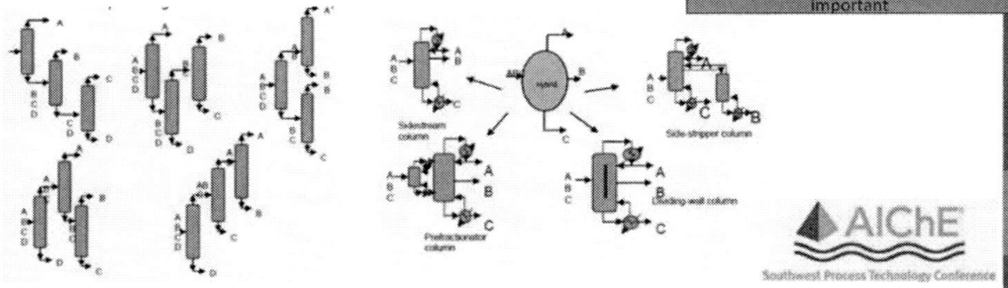

AIChE Southwest Process Technology Conference

Transformation of Hybrid

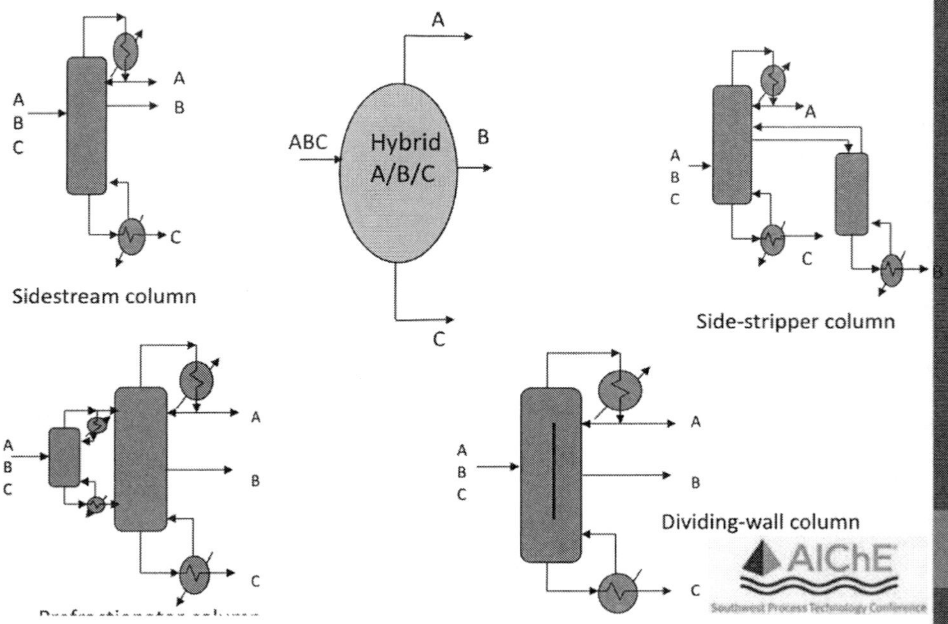

AIChE Southwest Process Technology Conference

8th AIChE Southwest Process Technology Conference I October 6-7, 2016 I Galveston, TX

Energy Savings in Separations

- Dividing Wall Columns
  - Value: 30% less capital and energy where properly employed
  - Potential Impact: Distillation accounts for 70% of the capital and energy of our traditional processes
  - Status: Full numerical and experimental validation available
  - Hurdle: Seen as an increased risk by Manufacturing ... reduced flexibility and control
    - Reality: BASF has installed ~80 of these units!
    - Exxon has operational units
    - Dow has a number of operation units
  - Key Energy Note
    - DWC will be better from a First Law perspective
    - Not necessarily better from a Second Law perspective

8th AIChE Southwest Process Technology Conference I October 6-7, 2016 I Galveston, TX

Dividing Wall Column Basics

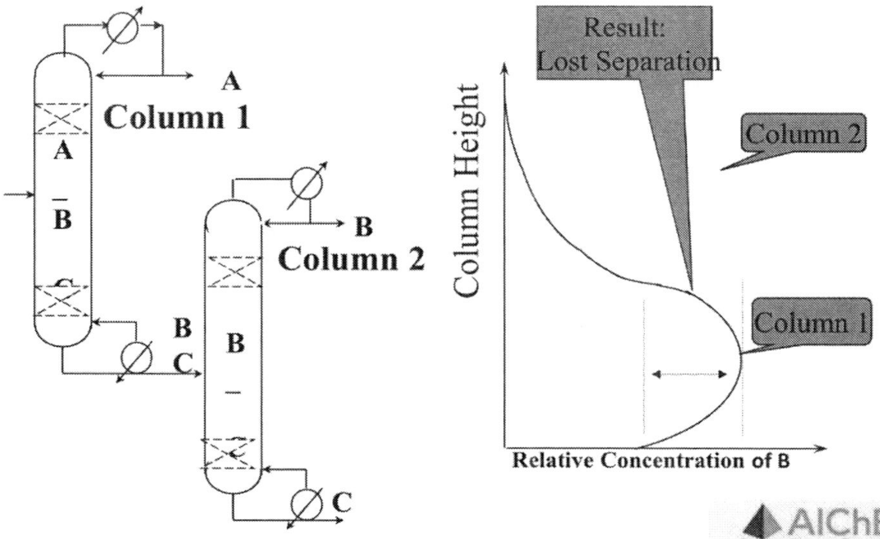

8th AIChE Southwest Process Technology Conference I October 6-7, 2016 I Galveston, TX

Dividing Wall Column: First Step

Fully Thermally Coupled Petlyuk Column

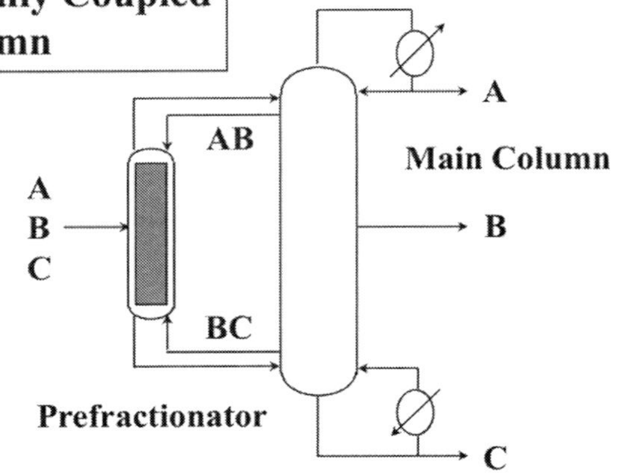

8th AIChE Southwest Process Technology Conference I October 6-7, 2016 I Galveston, TX

Simplified Schematic of DWC and Rigorous Simulation

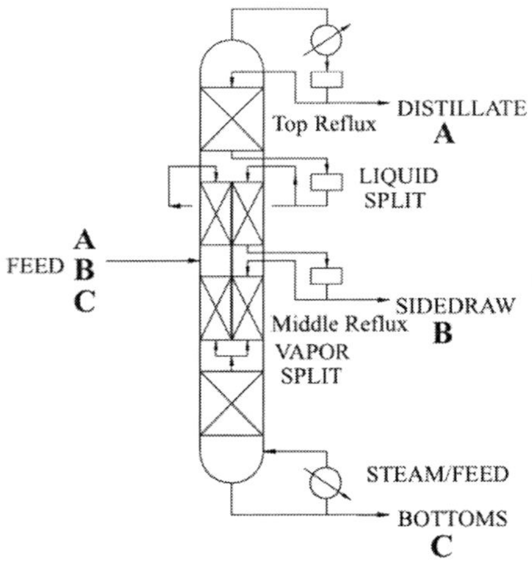

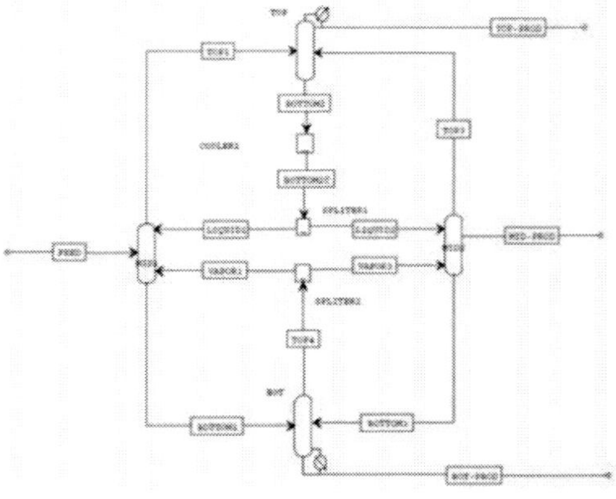

8th AIChE Southwest Process Technology Conference I October 6-7, 2016 I Galveston, TX

Energy Saving Arrangements Retrofit Opportunities

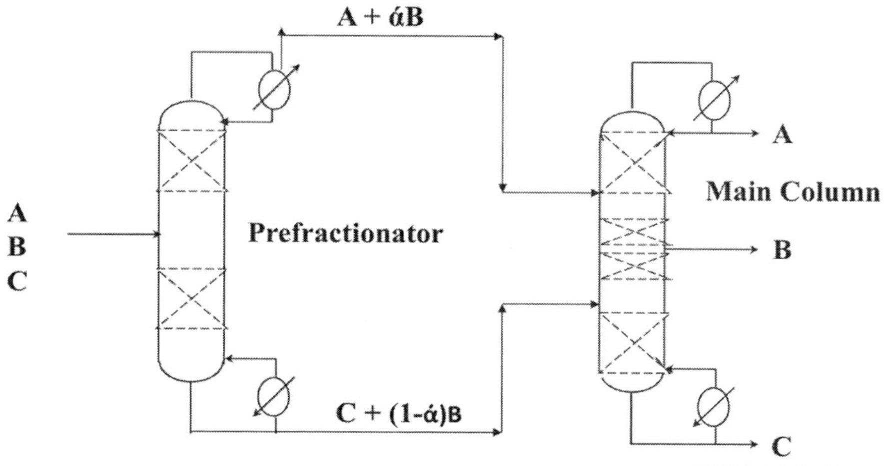

Energy Saving Arrangements

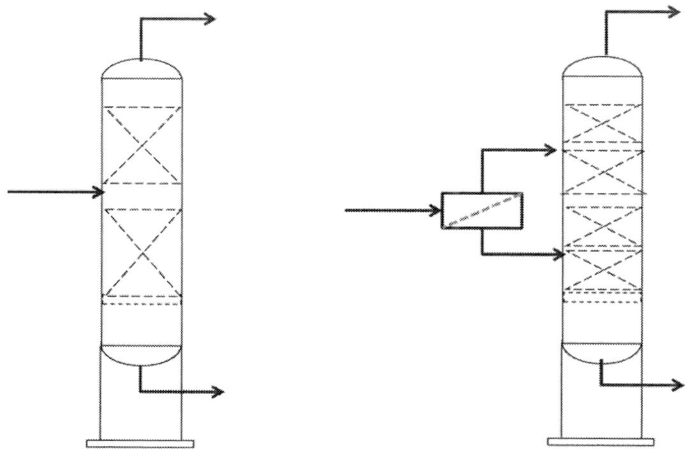

Distillation & Pre-fractionation Membrane

Energy Saving Arrangements

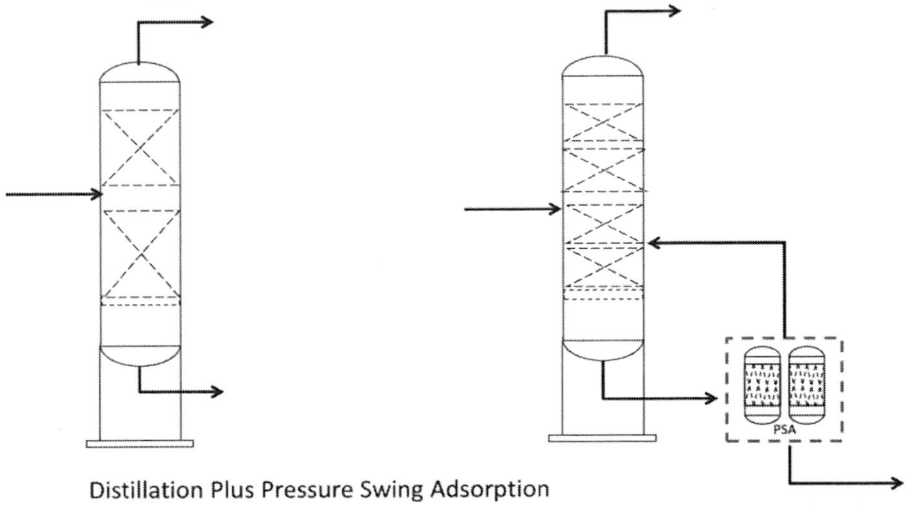

Distillation Plus Pressure Swing Adsorption

Energy Saving Arrangements

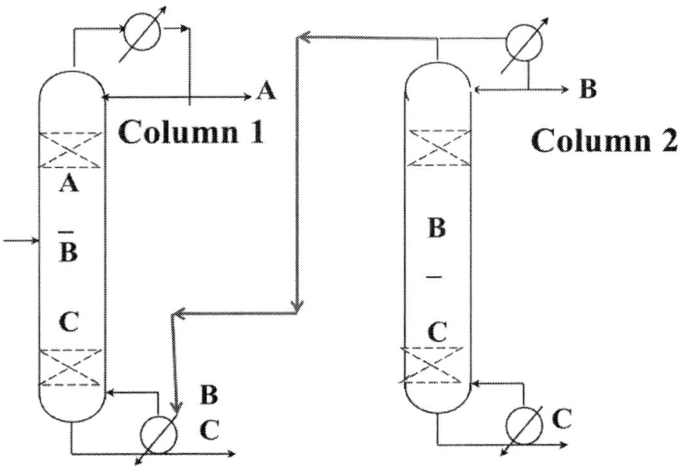

Heat Integration via Cross Exchange

Example: Large Scale Prefractionator

Patent US8129436 B2

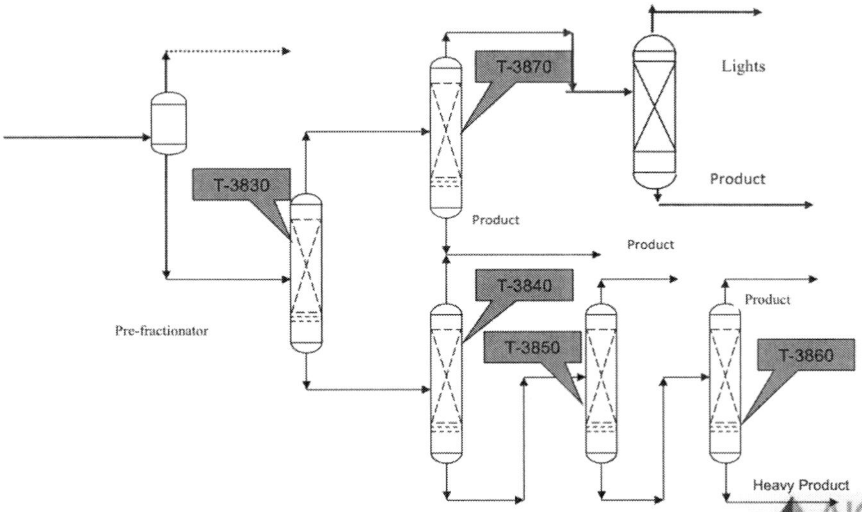

8th AIChE Southwest Process Technology Conference I October 6-7, 2016 I Galveston, TX

Example: Consolidation and DWC

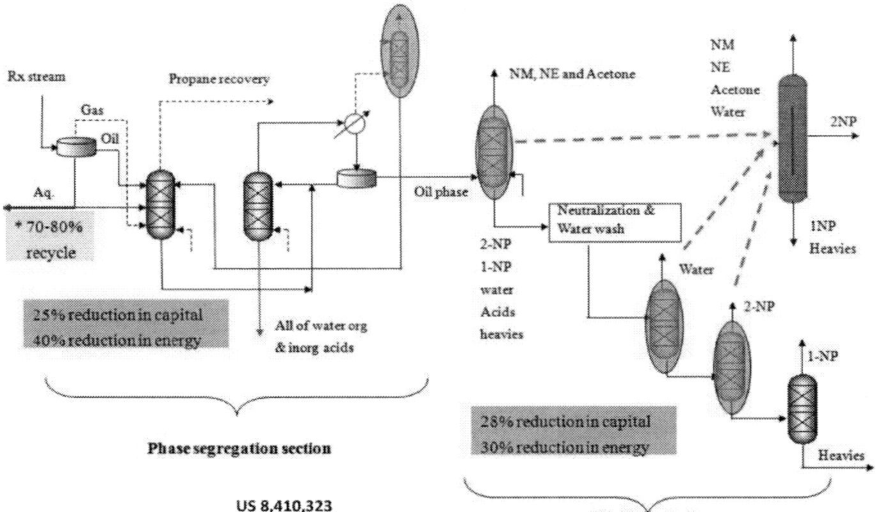

US 8,410,323
Process for Downstream Recovery of Nitroalkane using Dividing Wall Column

US 8,431,754: Process for nitroalkane recovery by aqueous phase recycle to nitration reactor

8th AIChE Southwest Process Technology Conference I October 6-7, 2016 I Galveston, TX

What is on the Horizon?

- Elements on the horizon that may alter the distillation research perspective
  - Legislation to lower greenhouse gases
  - Carbon taxes
  - Substantially higher energy costs
  - Mandates to lower potentially harmful inventories
    - Legislation
    - Community mandate

AIChE Southwest Process Technology Conference

8th AIChE Southwest Process Technology Conference I October 6-7, 2016 I Galveston, TX

Potential Game Changers

- Ionic liquids
  - Drawback: expensive, some have toxicity issues
  - Opportunities: can be highly selective, phase change
- MOF's
  - Drawback: expensive, not available in commercial quantities
  - Opportunity: can be highly selective
- Membrane reactors
  - Opportunity: Remove products, drive conversion
  - Drawback: temperature limitations and cost
- Reactive absorption
- Reactive PSA

AIChE Southwest Process Technology Conference

8th AIChE Southwest Process Technology Conference I October 6-7, 2016 I Galveston, TX

Thank You

- Dans ses écrits, un sage Italien, Dit que le mieux est l'ennemi du bien
- In his writings, an Italian wise (man) said that the perfect is the enemy of the good
- *Voltaire*

- Questions

Antoine
Technical
Consulting LLC

Biological Control of Water

Presented by **Keisha Antoine, PhD, PE** at AIChE
Southwest Process Technology Conference
Moody Gardens & Conference Center
October 2016

Overview

- About Me
- Use of water in industrial processes
- Biofilms and process issues
- Chemical treatment – disinfection & corrosion inhibition
- Summary

Keisha Antoine, PhD, PE
About Me

Education
B.S. Chemical Engineering, **M. Eng.** & **Ph.D.** Materials Science & Engineering, Lehigh University
Dissertation: "*In situ* investigation of photoinduced effects in arsenic-selenium glass films by x-ray photoelectron spectroscopy (XPS) and optical spectroscopy".

Career – 10 years in Industry
Antoine Technical Consulting, LLC, Principal/Process Engineer Consultant
Corning Incorporated, Senior Chemical Engineer, Chemical Process Engineer, Development Scientist

Other Qualifications
- Professional Engineer
- Peer reviewer, Journal of Non-Crystalline Solids
- National Science Foundation (NSF) Panel Reviewer
- 1 patent, 1 patent application
- Certified Green Belt
- 8 Publications, 11 presentations, 2 invited talks

Goal

- To provide an overview of chemical methods of control of biological contaminants in industrial process water
- We will discuss chemical and physical methods of biological control
- This is NOT a talk on waste water processing

Water in process use

- Water has a high heat capacity, i.e., it absorbs more heat for a given temperature rise than any other common inorganic substance. The steam is capable of carrying large quantities of heat ➔ heating and power generation
- What's in water? Dissolved gases, ions, suspended solids and other contaminants. Before process use, it must first be treated to an acceptable water quality standard appropriate for its final use
- Water impurities can cause problems with equipment leading to issues like reduced flow, high back pressure, reduced heat transfer, higher utility bills for pumping and heating, unexpected downtime from equipment failure

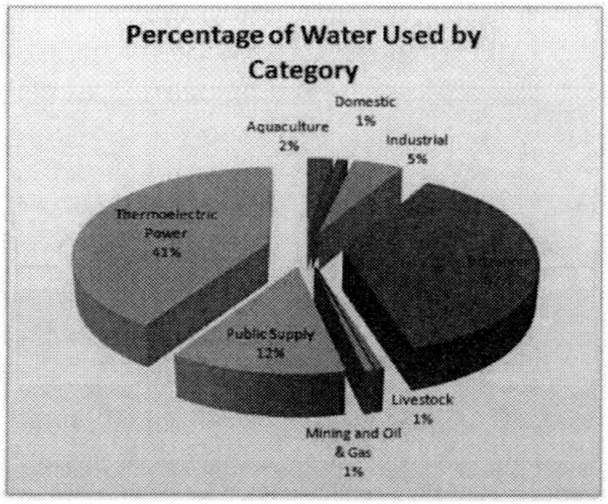

https://fracfocus.org/water-protection/hydraulic-fracturing-usage

Water Contaminants

Organic	Biological	Inorganic minerals	Radioactive minerals
Carbon-based – synthetic + natural organic matter (NOM). Synthetic: derived from petroleum e.g., dioxin, PCBs NOM: Humic acid, fulvic acid, amines, urea	Contact with the environment, air, humans and other animals Microbes Algae, Protozoa, Bacteria, Viruses	Minerals and toxic metals. Nutrients from Agriculture runoff From water source: Calcium, magnesium, iron From piping networks: copper, lead, arsenic, iron	Uranium, plutonium, radium A concern of produced water from fracking

High Level Process of Water Flow

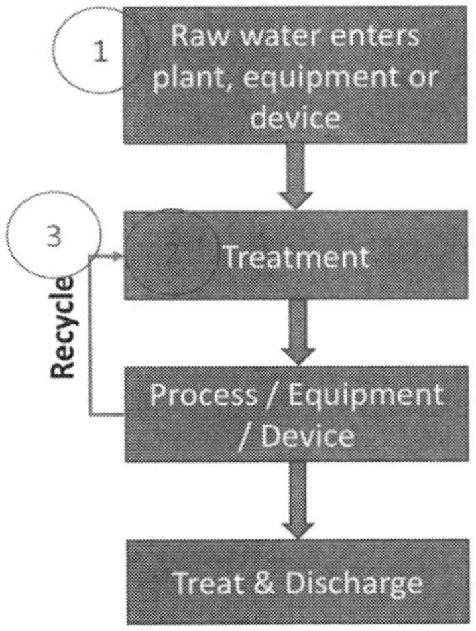

Process Flow	Microbes path of entry
Water source: River, Lake, Groundwater, Seawater, Municipal	✓
Provide water of a given spec eliminating or reducing: Suspended solids, Dissolved ions, Organic contaminants, Biological contaminants	✓
Water used for: Cooling, Boilers/power generation, Washing, rinsing, Incorporated in food/beverage	✓
After process use, a portion of the water is recycled and the rest is treated and discharged	

October 7, 2016

The goal of Industrial water treatment:
make efficient or optimize industrial water use

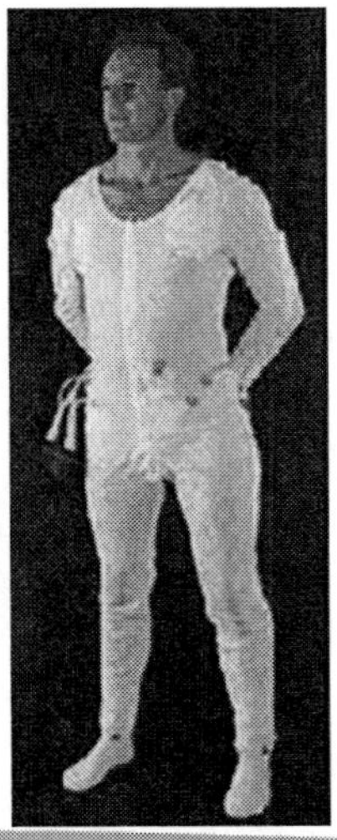

Heating & cooling

October 7, 2016

The goal of Industrial water treatment:
make efficient or optimize industrial water use

Storage

The goal of Industrial water treatment:
make efficient or optimize industrial water use

Processing

Fire suppression

We will focus on Biological contaminants in process water – not wastewater treatment

- Biological contaminants can lead to formation of biofilms
  - Reduced hydraulic diameter, Reduced flow, Reduced heat transfer, overheating of pipes and/or increased fuel usage. **BOILERS**
  - Provision of nutrients and protection for breeding of microorganisms that can be harmful to human health, e.g. Legionnaire's disease. **COOLING TOWERS.**
  - Sites for microbiological induced corrosion, leading to leaks and tuberculation and therefore shortened equipment lifespan. **ALL PROCESS EQUIPMENT.**

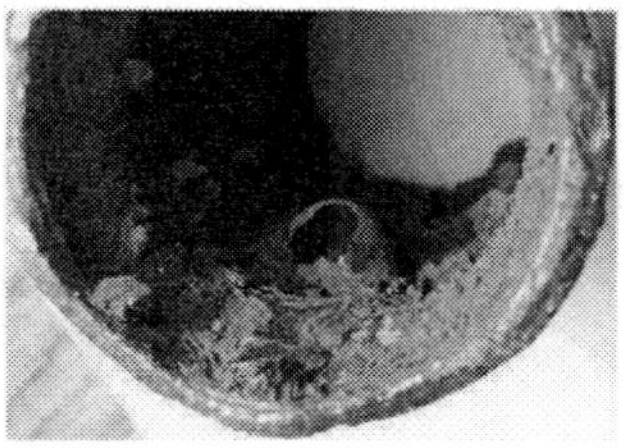

EXHIBIT 53.5 Interior Tubercle.

From Automatic Sprinkler Systems Handbook 2007, Supplement 3, Microbiologically Influenced Corrosion in Fire Sprinkler Systems, Bruce H. Clarke, Anthony M. Aguilera, p. 955

http://www.waterandhealth.org/battling-biofilms-aging-water-infrastructure/

Biofilm formation

1. Planktonic (mobile) bacteria introduced into water system through the raw water source, air, makeup water or process contamination

Biocides are most effective at Stage 1!

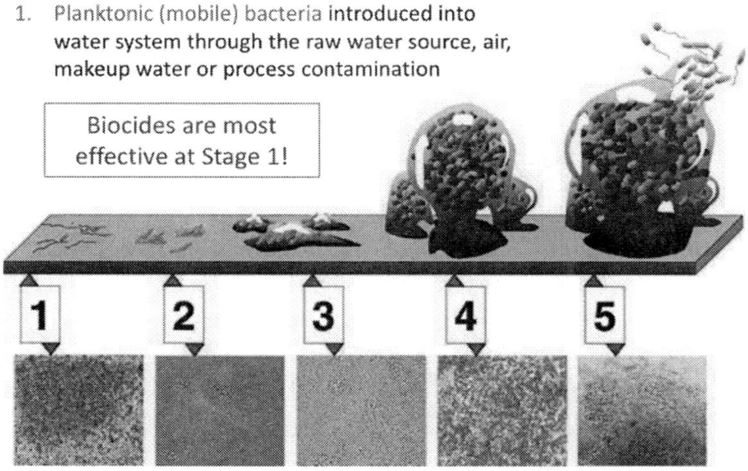

2 - 4. The bacteria become sessile (fixed to a surface) where they can grow and multiply, creating a biofilm.

5. Dispersion. The biofilm continues to grow until some of it disperses to colonize new surfaces. The process is then repeated.

By D. Davis - From: Looking for Chinks in the Armor of Bacterial Biofilms Monroe D PLoS Biology Vol. 5, No. 11, e307 doi:10.1371/journal.pbio.0050307 http://biology.plosjournals.org/perlserv/?request=slideshow&type=figure&doi=10.1371/journal.pbio.0050307&id=89595, CC BY 2.5, https://commons.wikimedia.org/w/index.php?curid=3364284

Mitigate biofilm formation by **Disinfection**

Disinfection – Keep the System Clean!

- Specialized cleaning techniques that destroy and prevent growth or organisms capable of infection.
- Expose microorganisms to chemical or physical agents

The effectiveness of disinfection is determined by testing for an indicator organism (total coliform bacteria). Although this organism is considered harmless, its presence indicates that pathogens may also have survived.

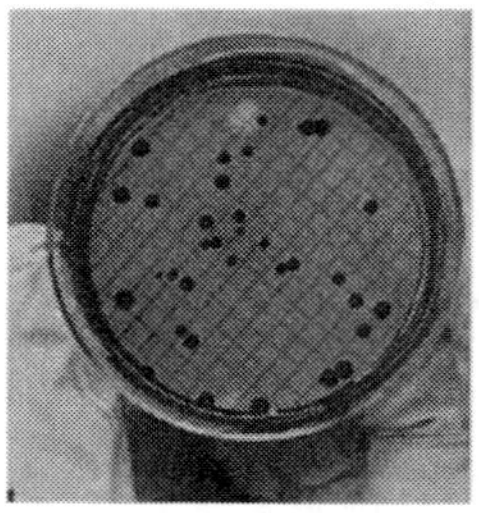

Image: *"Bacteria and Private Wells, Information Every Well Owner Should Know"*, Cliff Tyrens, Director of Public Awareness, National Ground Water Association

Disinfection Methods

UV Disinfection system by TrojanUV

Primary

- Chlorination
- Chloramines
- Ozone
- Ultraviolet light (UV)

Other

- Chlorine dioxide,
- Potassium permanganate
- Nanofiltration.

On-site sodium hypochlorite generation (chlorination) by De Nora

Considerations for picking a biocide

- Water chemistry (dissolved solids, particles of organic matter other non-pathogenic microorganisms)
- pH
- Type of microbes present and efficacy against those microbes
- Compatibility with corrosion inhibitors
- Retention time (length of time it remains active in the system)
- Contact time (CT) = disinfectant residual concentration [mg/L] x minimum contact time [minutes] of disinfectant with microorganism
- System volume
- Temperature
- Discharge points and permitting requirements
- Compatibility with materials of construction
- Cost

Types of Biocides

Oxidizing biocides – for maintenance
Chemicals that kill microorganisms by the electrochemical process of oxidation. The microorganism is oxidized by the oxidizing agent which is itself reduced by gaining electrons. Can create disinfectant byproducts, unwanted and oftentimes harmful if above defined thresholds. Continuous or shot-fed application.
Non-oxidizing biocides
They work through various processes, e.g., interfering with reproduction, stopping respiration, lysis. Generally shot fed to achieve a high enough concentration for a sufficient period of time (several hours up to a day) to kill the bacteria, algae or fungi.
Surfactants, biodispersants, biopenetrants
Chemicals that can penetrate and loosen the complex matrix of biofilms allow biocides to reach the organisms for more effective kill and control. Usually shot fed at dosages that break down polysaccharides, emulsify oils, release minerals and foulants or disperse the bipolymers

Be aware of **Disinfectant Demand**

Disinfectant demand

reactions of the disinfectant with substances other than pathogenic microorganisms

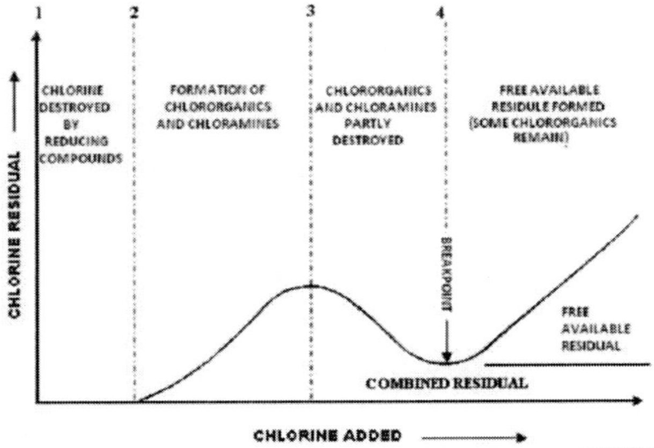

- Disinfectant demand consumes the disinfectant.
- Therefore, excess disinfectant must be added to provide a sufficient concentration of residual disinfectant to effectively kill pathogenic organisms.

Corrosion Inhibitors reduce Disinfectant Demand

https://www.nachi.org/forum/f22/copper-pipe-flux-corrosion-89835/

- Corrosion inhibition is especially important for unlined cast iron and unlined ductile iron pipes (old infrastructure). Ferrous compounds provide nutrients for microbial contaminants.
- Although corrosion inhibitors are phosphorous-based, which itself may be a nutrient source, these inhibitors:
  - ↓ the leaching of lead and copper ➔ ↓ disinfectant load and ↑ residual disinfectant

Effective corrosion control + disinfection = ↓biofilm growth + ↓microbiological contamination

Summary

- **Keep system clean! Reduce available nutrients & disinfectant demand**
  1. At design stage, incorporate use of alternative piping materials to unlined cast iron
  2. If older system using unlined cast iron, ensure an effective anti-corrosion program using phosphate-based inhibitors.
  3. Maintain high residual chlorine (within permitted limits) or other oxidizer for secondary disinfection. High residuals can be a deterrent to biofilm formation
- **Choice of treatment plan depends on system**
  - New vs. in-service and fouled
  - pH and other process conditions
  - Permitting requirements/limits

The Latest Advancements in Electric Heating and Control Technologies for Chemical Processing Applications

Taylor, D., Chromalox

Originally presented on: 10/7/2016 14:30:00 - 15:00:00

Chemical processing industries have historically been dependent on large fossil fuel fired heating systems, which are piped throughout processing areas to provide heat for multiple users and applications in a given plant. Utilizing electric resistance heating for chemical market applications is not a new concept. However, recent advancements in electric heater and control technologies are expanding the role and number of heating applications capable of utilizing electricity for process heating. It has become the preferred method for process heating and temperature maintenance for many applications due to both performance and financial reasons.

The benefit of using electricity instead of fossil fuels extends beyond the localized zero emissions associated with electric process heating. This presentation walks the audience through all the features and benefits of utilizing electricity instead of the traditional fossil fuel fired heating systems. Specifically, new technology advancements in medium voltage heating and control systems, high temperature heat tracing, and wireless instrumentation have iopened the door for Chemical facilities to leverage electric for their process heating and temperature management requirements.

Not only will the presentation explain these new technologies, but it will also highlight chemical processing applications where they have been used effectively. Finally, case studies will be presented where electric heating and control systems have been successfully implemented and achieved stated performance or financial benefits over fuel fired systems.

AUTHOR INDEX